BEAT CONSTIPATION
Without
LAXATIVES
AND
LOSE WEIGHT
THAT IS SUSTAINABLE
and PERMANENT

BEAT CONSTIPATION
Without
LAXATIVES
AND
LOSE WEIGHT
THAT IS SUSTAINABLE
and PERMANENT

USEN IKIDDE

authorHOUSE®

AuthorHouse™ UK
1663 Liberty Drive
Bloomington, IN 47403 USA
www.authorhouse.co.uk
Phone: 0800.197.4150

Published by AuthorHouse   10/23/2014

ISBN: 978-1-4969-9294-9 (sc)
ISBN: 978-1-4969-9295-6 (e)

Library of Congress Control Number: 2014917981

<u>If you have not read the first and last five chapters of this book, then you have not read this book.</u>

THIS BOOK IS A REVELATION.

IT ESTABLISHES A RELATIONSHIP BETWEEN
CONSTIPATION AND-

DIABETES
ISCHAEMIC HEART DISEASE
STROKES
CANCER
HYPERTENSION
OBESITY

(DISCHO)

AND HOW THE "O" IN "DISCHO" CAN BE REVERSED INTO <u>WEIGHT</u> <u>LOSS</u> THAT IS SUSTAINABLE AND PERMANENT.

IT ALSO ESTABLISHES A RELATIONSHIP BETWEEN CONSTIPATION, THE ABOVE CONDITIONS, AND CIGARETTE SMOKING, EXCESS SALT, SUGAR AND ALCOHOL. BE AMAZED. READ ON !

USEN IKIDDE

ABOUT THE BOOK

"Beat Constipation Without Laxatives And Lose Weight That Is Sustainable And Permanent" is essentially for lay people. It is a compelling read which defines constipation in terms of the most common presentations of Straining, Hard stool, Infrequency, Insufficiency, and Difficulty (SHIID). It sets out the prevention of constipation by adequate 3FEV factors (Fruits, Fibre, Fluids, Exercise, Vegetables) - the bedrock.

The relationship with the serious conditions of Diabetes, Ischemic heart disease, Strokes, Cancer, Hypertension, Obesity (DISCHO) is established in a manner that is compelling, unexpected and fascinating.

The fight against DISCHO is a fight against Cigarette smoking, reduction in Salt, Sugar and Alcohol consumption, (CISSA) is another powerful relationship with DISCHO and Constipation.

Very importantly the "O" in DISCHO is reversed and mobilized as a means of WEIGHT LOSS that is sustainable and permanent and spelled out in a manner never before expounded by exploring BMI and also calories in various food items and utilizing these and exercise effectively for permanent weight loss.

Lastly important factors are clearly crystallized in the principle of gastro colic reflex plus the embodiment of positions to achieve better bowel evacuation and spot exercise - the ultimate mobilization to fight constipation which ends this long journey of discovery.

PREFACE

This book is intended primarily for the layman. It is useful for all ages especially those who constantly experience constipation. It would be rare to meet anyone who has never experienced constipation or unlikely to experience it at some stage in their lives. This book is therefore a sine qua non for everyone. The last chapter is the new, most revealing and very useful information which could transform your life and provide an avenue to deal effectively with constipation and get immediate result without the need for laxatives and suppositories. It is especially a must for the over 50s. For the first time it shows how a simple, short exercise combined with other measures can be used to overcome constipation and yield immediate result which can be used whenever constipation recurs thereby obviating the need for the widespread use of laxatives which can cause unpleasant and serious medical consequences. Indeed **"exercise adds years to your life and life to your years" and can also take away constipation from those years**. Chapter 10 mobilises the weapons in the fight against constipation to combat **obesity** and achieve **weight loss** that is **sustainable** and **permanent** The medically minded reader may also find it useful as it deals with some of the patho- physiology of constipation, management, medical causes and parts of the gastrointestinal tract that are mainly concerned with constipation.

Constipation is the passage of hard stool which is less frequent than your normal bowel habit. Constipation is a relative term as it means different things to different people. To some people it means the passage of hard stool which is less frequent than the normal bowel habit of the individual. Others consider other elements such as the difficulty and straining. The common denominator in constipation are the following: - **S**training, **H**ard stool, **I**nfrequency, **I**nsufficiency and **D**ifficulty **(SHIID).** This is elaborated in this book.

It is important to stress from the outset that there are also medical causes of constipation which you should exclude by visiting your family doctor/GP before embarking on this very simple exercise in chapter 12. I must equally stress that normal bowel habit varies between individuals from about once to thrice a day to even once to thrice a week in some people. In fact, in extreme cases there are a few "normal" people who have been known to only open their bowels once in a few weeks. A change in bowel habit must always be investigated by visiting your family doctor.

A simple non strenuous exercise lasting an average of 20-50 minutes which could work virtually all the time if combined with other measures is a revolutionary step which obviates the need to take laxatives which can be "habit forming" in the sense that there are many people especially the elderly who take laxatives or faecal softeners on regular basis before they are able to open their bowels. This has far reaching medical consequences especially low potassium which in extreme cases could be fatal.

I shall use the last chapter of this book and take you through this simple exercise which I believe many people would find so useful that they would begin to wonder why they did not know it over the years. The truth is that no one knew or thought about it. Although it has long been known that exercise in a general sense improves bowel functioning and helps to prevent constipation. This particular exercise which is more specific should generally give immediate results if properly performed.

The initial purpose of this book is to explain how this simple exercise is performed. I will of course, look at other important aspects of constipation, in particular other forms of management and those that should be combined with the exercise routine to expidite and improve the desired outcome. There are other very important measures such as the 3FEV, explained in this book, which should be combined with the exercise. The exercise is therefore not the only measure advocated.

There is something unusual but quite specific which has come out of this book: **that there is a relationship between constipation and the global "epidemics" of myocardial infarction (heart attack), hypertension, strokes, diabetes, cancer and obesity**. That relationship is the commonality in the instruments of the fight against these serious and life-threatening conditions and constipation. These are emphasised in various sections of this book.

The principles of **3FEV** and **CISSA** embodied in this book and the relationship with **DISCHO** with **O=OBESITY** is the most natural means of **WEIGHT LOSS** to **burn your calories** and ensure that the **weight loss stays permanently.**

You cannot afford to miss it; read on!

This information is a sine qua none globally but **specifically for the industrialized and developing nations adopting European/ Western life style especially the United States Of America, United Kingdom, Russia, China, Nigeria, Germany, France, Australia, Canada, Spain, India, Japan, Israel, Italy, Indonesia, Brazil, Argentina, Sweden, Norway, Denmark, Finland, South Africa, Portugal, Egypt, Ghana, Kenya, Ethiopia, Sudan, Mexico, Jamaica, Pakistan, Turkey, Saudi Arabia, Chile, Venezuela, Iran, Zambia, Republic Of Ireland, Emirates, New Zealand, Korea etc**

To the layman/woman to which this book is intended, some chapters are to be "tasted only, others to be well masticated, digested, absorbed and assimilated" The latter applies to the first and last five chapters of this book.

I would therefore like to re-emphasize that **if you have not read the first and last five chapters of this book, then you have not read this book.** Don't stop here; read on and be amazed!

CONTENTS

CHAPTER ONE

What is Constipation and What Causes It?

Constipation is the passage of hard stool which is less frequent than the individual's normal bowel habit. This definition of constipation is simplistic because different people look at the symptoms of constipation from different angles. To some people constipation means the passage of hard stool irrespective of the frequency. Others view constipation as infrequency in bowel movement, irrespective of the consistency and there are others who regard constipation as difficulty in the actual act of expulsion of stool; this is often due to hard consistency but may be due to other causes to be discussed later. To some people, it's insufficiency in the amount of stool that constitutes constipation but this is more likely due to hardness or lack of bulk or in a minority of cases to obstruction.

To understand these symptoms- complex of constipation, we shall look at the common denominators which constitute the main complaints associated with constipation. I shall state them as follows-

Straining
Hardness of stool (consistency)
Infrequency
Insufficiency
Difficulty

This is best remembered as **SHIID** of constipation. Straining is a common complaint and is usually due to hard stool whilst hard stool is the result of excessive water absorption due to the slowing down of colonic contents. The slower the journey, the more the water absorption is likely to be and the worse the constipation. Infrequency is probably the hallmark of constipation: most people first become aware that they are constipated when they have not had their usual bowel motion at their usual times. This is almost always accompanied by hardness of the stool. The two tend to go hand in hand and are probably the two most common complaints in constipation. Insufficiency (small amount of stool) is a result of inadequate bulk of stool due to lack of adequate fibre content, decreased motility and excessive water absorption. Difficulty in passage of stool is usually due to hardness of stool which also causes straining as noted above.

In a "normal" constipation not caused by intrinsic or extrinsic factors, it can not be emphasised enough that constipation (or SHIID) is due to inadequate intake of fruits, fibre, fluids, exercise (participation) and vegetables which is shown below and constitutes FFFEV or 3FEV. Adequate 3FEV is the corner stone in the fight against constipation.

Unfortunately, to some people the mere mention of the word constipation is still a taboo to be avoided. It falls within the realms of privacy; not to be discussed in public. Such notion is very unfortunate because almost everyone has either had the symptoms of constipation, has it now or will have it in the future. Therefore you should be able to discuss it freely with your family doctor, General Practitioner (GP) or any healthcare practitioner or professional. The building blocks in the foundation for the prevention of constipation and indeed for a healthy lifestyle are listed below-.

Fruits
Fibre
Fluids
Exercise
Vegetables

A proper understanding and utilization of the **3FEV** above also happens to be one of the greatest weapons in the prevention and fight against heart disease, strokes, hypertension, cancer, obesity and diabetes. In a nutshell, the leader in this fight is exercise since on average most people "inadvertently" have some fruits, fibre and vegetables of some sort in their diet without much thought of their importance. Exercise is important because there are several people who don't exercise at all. By sheer coincidence this happens also to be the main theme of this book which is contained in the last chapter. Thus, exercise is not only an essential weapon against heart disease, strokes, hypertension, cancer, diabetes and obesity but also **constipation**. That is where the commonality lies-the similarity in the instruments of the fight against these conditions.

To the medically minded, **3FEV** can easily be remembered as 3 **F**orced **E**xpiratory **V**olumes. which of course has nothing to do with forced expiratory volume but everything to do with the 3 FEV of healthy living above.

By far the largest number of my intended readership are lay people and for whom this book is primarily intended and will remember this mnemonic as **3F**emale **E**xamples of **V**irtue. These virtues are those of healthy living by eating adequate portions of fruits, fibre and vegetables, drinking adequate fluids and having a good exercise routine. These are the virtues here.

To make sure that these virtues do not end on the pages of this book but as a national and international campaign of healthy living, the question should be asked: **What are the 3 Female Examples of Virtue ?** This question should be asked **everywhere, in family circles at home, to friends, in school, on playground, in the pub, dinner table, parliament, internet friends in face book, twitter etc and test knowledge of the responders-** fruits, fibre, fluid, exercise and vegetables. Surprise your friends, mum, dad, wife, husband daughter, son, boyfriend, girlfriend and acquaintances by giving them a copy of this book. In short, spread the message of 3FEV. It is probably the greatest weapon in the fight against ill health and applies equally to both sexes as well as to everyone irrespective of age.

This is an example when 3=5 because there are actually 5 examples in the real sense and 3 is just part of the mnemonic representing the 3Fs - fruits, fibre and fluids and the other 2 factors are exercise and vegetables. Some boys may object to an only female connotation and may prefer to ask the question: what are the 3 **F**amous **E**xamples of **V**alour? The answers are the same. Some ladies may also prefer to ask the same question.

By so doing you would be helping to propagate this message of healthy living amongst friends, family, colleagues and even people you have never met and don't hesitate to give them the answers or where to find them and test them again to see whether they have got the message. Make it fun and quiz to be enjoyed!

It is hoped that this message will percolate and spread like wild fire throughout the globe and to all countries including the **United States Of America, United Kingdom, China, Russia, Canada, Nigeria, India, Germany, France, Brazil, Argentina, South Africa, Australia, Scandinavian countries, Indonesia, Spain, Israel, Portugal, Egypt, Ghana, Kenya, Italy, Mexico**, **Japan, Jamaica, Pakistan, Iran, Emirates etc**.

I have mentioned certain countries by name, perhaps western and emerging economies where lifestyle issues cause particular problems or are just around the corner especially the problem of obesity which has reached "epidemic" proportions in some countries, but the list is by no means exhaustive, nor are the problems peculiar to these countries only. The problems of heart disease, strokes, hypertension, diabetes, cancer and obesity are global and the adoption of western lifestyle in emerging and developing economies has seen the "surfacing" of hitherto unknown conditions or diseases.

The simple question: "What are the 3 female examples of virtue?" provides the first set of weapons in the fight against heart disease including myocardiac infarction (heart attack), hypertension, strokes, obesity, diabetes and cancer and, of course, constipation.

I hope to dwell more comprehensively on the fight against these diseases in future publications to avoid digression from the theme of this book. However, it is important to point out the similarity in the weapons in the fight against constipation and the serious health problems. The similarity was simply unexpected but the reality is that the commonality exists and should be looked at very carefully as a means of "killing two birds with one stone" and as we shall see later, not just 2 birds but 3.

Thus there are 5 main symptoms of constipation, namely; straining, hardness of stool, infrequency, insufficiency and difficulty (SHIID). The common pathway for constipation seems to be the consistency - hardness of the stool which causes straining and reduces frequency or causes difficulty in the passage of stools and inadequacy in the amount. Stool becomes hard when the passage of stool through the large bowel (colon) slows down leading to excessive absorption of water by the gut. There is also no easy definition of normal bowel habit: this is influenced by diet, custom and individual habit. There is therefore a wide variety of "normal", varying from 1-3 times a day to 1-3 times a week. Infact, there are few people who open their bowels only once a week and is considered normal for them. It is most important to restate once again that a change in one's normal bowel habit is important. Under such circumstances it is important to consult your doctor or GP before embarking on the simple exercise routine which is described in the last chapter of this book. We can now look at the causes of slowing down of the stool in the large intestine.

There are many reasons for this but one of the commonest cause is lack of physical activity especially if other causes are eliminated. That is why this book is primarily devoted to increasing physical activity in a regular and rhythmic pattern. Increase in physical activity helps to improve regular inherent bowel contraction (peristalsis) which results in hastening or propelling colonic contents along the large intestine resulting in less absorption of water thereby avoiding constipation.

People who are physically active experience less constipation than physically inactive people. Patients who are bed-ridden for long periods of time have a higher risk of experiencing constipation. Constipation is more problematic in the elderly than the young who are more physically active whereas the elderly tend to have a more sedentary life-style than the young who are usually more active and out- going. The young quite often engage in sports like athletics, football, cricket, cross country races, rugby, tennis and basket ball which the elderly can only dream of. There is also the general slowing down of metabolism and general and natural body mechanisms. The elderly are also more likely to be on several medications-some of which cause constipation as will be seen later.

There are, of course, other very important causes of constipation which will be dealt with in other chapters in this book. These include poor diet lacking in fibre, drinking inadequate fluids, diets lacking in fruits and vegetables, medical causes especially cancer leading to obstructive symptoms, neurological disorders-strokes, multiple sclerosis (MS), irritable bowel syndrome, Parkinson's disease, injuries of the spinal cord and pseudo obstruction of the intestine (unknown cause).

There are systemic diseases which affect several body organs and tissues which can cause constipation. These may also affect the intestine and slow down normal contraction of the intestine (peristalsis) and thereby causing excessive absorption of water and hardening of the stool leading to constipation. Such diseases may include sclerodema, amyloidosis and lupus.

Excessive use of laxatives may have the opposite desired effect as a result of the habitual use which has to be satisfied in order to be able to open the bowel. This means that the bowel gets so used to having laxatives, it would not move unless laxatives, suppositories or enema are used. This may have several deleterious effects; namely habit formation and use of increasing doses before being able to open the bowel. There is also the more serious effect of excessive loss of potassium which can affect body tissues and organs and be fatal due to its effect on the heart.

Sometimes we actually cause or potentiate constipation by disobeying the call of nature by suppressing the urge. This usually happens because we are too busy at work, during travel when we feel that public toilets are not clean enough. In children, they are often too busy playing with their friends to care and in the elderly it may be quite a task getting out of bed due to pain or some prevailing medical conditions in which getting out of bed would cause pain. In some cases painful conditions of the anus or rectum such as piles, anal fissures and uncomfortable rashes around the anus discourage people from obeying the call of nature. This often leads to a vicious circle in which existing constipation becomes worse and the consequent hard stools become harder and worsen the existing condition. Under these circumstances, it is common to completely ignore the call of nature.

There are several medications which cause constipation. These will be dealt with in a later chapter. Suffice it however, to mention them briefly here. They include narcotics (Opiods especially morphine and codeine and codeine containing medication like Co-codamol) antidepressants like Amitrytiline and Imipramine, drugs used to control epilepsy (anticonvulsants) such as Carbamazepin (Tegretol). Iron tablets also cause constipation, some antacids, calcium blockers, milk and milk products are also known to cause constipation in some individuals.

Pregnancy may cause constipation due to hormonal changes in general but specifically due to the compressing effect of the foetus on the intestine which slows down the passage of food or faeces.

It is pertinent to note that most causes of constipation are due to problems in the eventual pathway in the gut which leads to slowing down of gut motility and excessive absorption of water. It is important to prevent these effects by adequate diet rich in high fibre, fruits and vegetables and regular exercises to help peristalsis and ensure fast movement of food and faecal matter and by so doing reduce excessive water absorption thereby preventing constipation and its consequences.

This book aims to emphasise how this can be done and above all for the first time show how a simple hitherto unknown exercise routine taking a few minutes can produce immediate results if properly performed and preceded by simple healthy habits which can be simplified, regularised and indeed habituated to become part of healthy lifestyle.

May I mention at the end of this chapter that the 3FEV are not the only weapons in the fight against **constipation** and the serious conditions of **D**iabetes, **I**schaemic heart disease, **S**trokes, **C**ancer, **H**ypertension and **Obesity (DISCHO),** other very important health measures (**CISSA**) are also involved. Please see chapter eight.

There is inter woven relationship between **CONSTIPATION, 3FEV, DISCHO, CISSA** and **WEIGHT LOSS** which is probably the most intriguing part of this book The intertwining in the relationship between constipation and the most important health and disease entities and their prevention is curious and unexpected. There is explanation later in this book. You can not afford to miss it.

Chapter Two

Functions of the Colon in Relation to Constipation

The digestive tract consists of different parts of a smooth muscular tube that squeezes food along, comparative to squeezing toothpaste from a tube.

The story would be incomplete without a brief mention of how the food gets to the colon. In reality the journey begins in the mouth. When we eat, the food is chewed or masticated by the teeth and we voluntarily swallow it, propelled by the tongue into the pharynx where the upper sphincter of the gullet relaxes and by peristaltic action controlled by the brain and mediated by autonomic nervous system, the food continues to the lower gullet where the sphincter relaxes to let it enter the stomach.

Digestion of food begins in the mouth when amylase present in the saliva starts the process of breaking down of the carbohydrates. Digestion is the process of breaking down the food into smaller components which can easily be absorbed and assimilated by the body.

Digestion continues in the stomach by the action of gastric juice - a combination of hydrochloric acid and pepsin which commences the digestion of protein. This process takes about 1-2 hours. Mucus secreted by the stomach helps to protect the lining of the stomach against the damaging effects of the hydrochloric acid. After some mixing of the food in the stomach, the chyme as it is now called enters the first part of the small intestine- the duodenum. The entrance to the duodenum is guarded by the pyloric sphincter, a valve which opens to let in the chyme. Here digestive enzymes are secreted from the pancreas (trypsin which digests protein, and pancreatic lipase digests fat usually initially in the inactive forms) and the release of bile from the liver helps in the emulsification of fat and the formation of fatty acids for absorption. There are other intestinal enzymes which help to complete the process of digestion in the small intestine and ensure their absorption and assimilation into the body. The enzyme lactase is involved in the digestion of carbohydrates by breaking down lactose into glucose and galactose. People who produce insufficient lactase are said to be "lactase intolerant" and can not deal adequately with lactose in milk.

Another enzyme- sucrase breaks down sucrose found in sugar into glucose and fructose which can be absorbed by the small intestine. The small intestine consists of three parts-duodenum, jejunum and ileum, all of which are involved in this process to varying degrees.

On completion of digestion and absorption of nutrients into the blood vessels and the lymphatic system of the small intestine, the indigestible material notably fibre derived mainly from fruit and vegetables or possibly some ingested de novo and artificial fibre now continue the journey into the large intestine.

LARGE INTESTINE

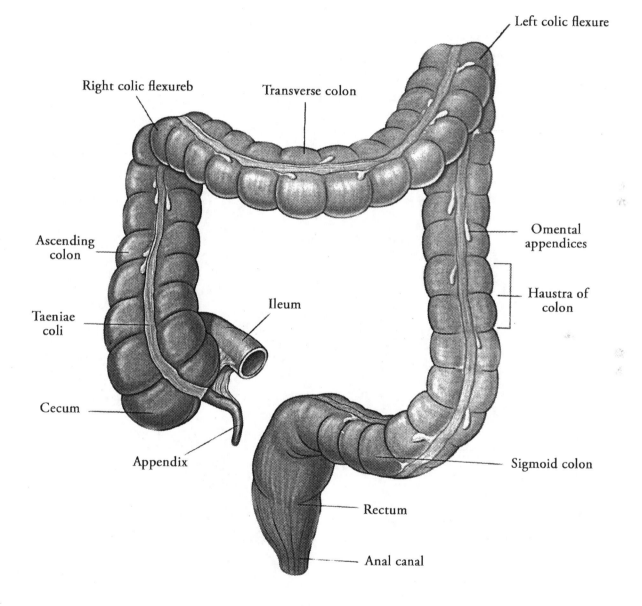

Left colic flexure

Right colic flexureb

Transverse colon

Ascending colon

Omental appendices

Taeniae coli

Haustra of colon

Ileum

Cecum

Appendix

Sigmoid colon

Rectum

Anal canal

With Kind Permission of ELSEVIER Churchill Livingstone.

RECTUM AND ANAL CANAL

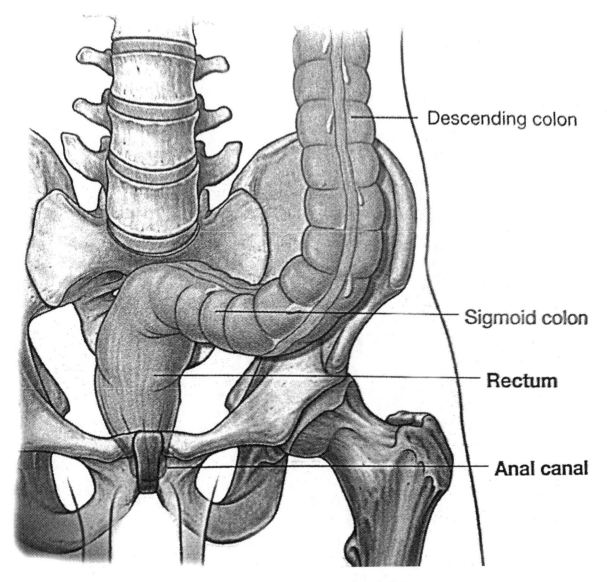

Descending colon

Sigmoid colon

Rectum

Anal canal

With Kind Permission of ELSEVIER Churchill Livingstone.

This book is concerned mostly with the last part of this journey in the colon down to its termination in the rectum and the anus. The large intestine (colon) as its name suggests has the largest diameter compared to other parts of the intestine. It begins at its largest part called the **caecum** at the termination of the small intestine which joins its lower part. The appendix also joins its lower posterior (back) part. The caecum is continuous upwards as the **ascending colon** where it lies just below the liver and appropriately named the hepatic flexure. From here the colon continues in a transverse course - the **transverse colon** to the left where it lies below the spleen and named the splenic flexure. The transverse colon is the longest part of the colon, and at the splenic flexure it continues downwards and to the left as the **descending colon** which at its lowest part assumes a somewhat S shape and appropriately named the **sigmoid colon** before continuing downwards as the **rectum**. The sigmoid colon is the narrowest part of the large intestine. The rectum consists essentially of 3 parts; the upper part that is continuous from the sigmoid colon, a middle part and lower part. The rectum increases in size as it descends downwards from the sigmoid colon until it joins the anus at its termination. Here the rectum assumes a considerable size in diameter and capable of greater expansion. The rectum is cylindrical in shape, not sacculated like the other parts of the colon. It is about 6-8 inches in length. The large intestine is about 5 feet in length and about one fifth of the length of the gastrointestinal tract. This is an adequate absorptive surface for water and a few other products which will be mentioned later.

Under normal circumstances about 1Litre of the contents of the last part of the small intestine (the ileum) containing about 90 percent of water are released into the first part of the colon (the caecum) each day in the adult. During its journey through the colon, absorption of this water takes place leaving only 100-200ml in the stool for excretion. About 70 percent of faeces is water and 30 percent is solid of which half is food waste and desquamated lining of the gut and the other half is mostly bacteria. Apart from the absorption of water, negligible amounts of vitamins, amino acids, fatty acids and glucose can also be absorbed by the colon.

There is an active absorption of sodium and passive diffusion of potassium back into the faeces. In diseased states of the colon, for example, where there is inadequate area for absorption, there may be a reversal or less absorption of sodium and excessive excretion of potassium into the faeces. As will be evident later in this book this situation can also occur with the use of some laxatives leading to excessive loss of potassium in the faeces which can have a deleterious effect on the functioning of vital organs particularly the heart. Normal potassium level is vital for the functioning of the heart. Inadequate blood level (hypokalaemia) can affect the heart adversely and cause cardiac arrest. The reverse is also true: excessive blood level of potassium (hyperkalaemia) can equally have the same effect on the heart. Usually the body maintains a balance - an equilibrium in which there is neither inadequate nor excessive levels of this element. Potassium is a vital element at cellular level without which the cell can not function effectively. Normally we maintain normal levels without having to worry about it, but in diseased states or interference as in excessive use of laxatives, excessive loss of sodium in the stool can occur as well as re absorption of potassium into the body. Sodium and potassium have an inverse or opposite relationship at cellular level: they somewhat avoid each other, when there is less sodium there is more potassium and vice versa. When laxatives are used there is a general loss of electrolytes including sodium and potassium without cellular consideration leading to depletion of electrolytes which with regular use without replenishment could spell dire consequences.

The fear of constipation in some individuals may lead to a number of symptoms including abdominal distension or fullness, bloating, heaviness, headache, nausea, anorexia, low mood, and coated tongue. They often attribute these symptoms to absorption of some toxic materials due to the constipation. There is no medical evidence for this assumption. Some people even refuse to eat and "load themselves" again when they cannot get rid of what they already have, others resort to laxatives which unfortunately become habitual culminating in a vicious cycle.

In general, motility of the large intestine is influenced by the food we eat, physical activity, faecal bulk, emotional stability or instability.

Colonic motility is a more complex phenomenon than simple peristalsis in other parts of the intestine.

There are 3 main types: -

1. Mass movement occurs in response to food particularly in the transverse and descending colon and possibly only occurs 1-3 times daily. This is a strong circular contraction which propels contents of the colon distally and over long distances.
2. Segmentation is again seen in the transverse and descending colon and generally divides the colonic lumen into segments as the name implies. Unlike mass movement, segmentation is more irregular and only occurs over shorter distances and in both directions. It is analogous to the movement of a larva of an insect.
3. Retrograde peristalsis is more often experimental than real. As the name implies the movement is retrograde moving upwards (descending colon and sigmoid colon) or backwards (ascending colon) and seen in descending, right and sigmoid parts of the colon. Its role in the propulsion of faecal matter is ineffective and doubtful.

Under normal circumstances, a meal would reach the caecum after 4 hours, and takes another 20 hours to reach the sigmoid colon and rectum. The flow of faecal matter appears to be rather irregular as recently eaten meal residue may actually bypass an already existing residue in the caecum. This phenomenon may be due to the type of meal residue and its consistency. For this reason it sometimes takes 3-4 days for some meal residue to be passed out. It is therefore very important that we pay particular attention to what we eat as it appears more than likely that certain food residues like fruit, vegetable, high fibre diet are more likely to make the journey faster than certain other items of food. and therefore the latter are more likely to lead to **constipation**. Secondly, the current emphasis on possible carcinogenic (cancer forming) nature of some food items suggest that the sooner we get rid of them naturally from our system, the better, and the longer they remain in the gastrointestinal tract the more carcinogenic they are likely to be due to increased contact time. We should, however, not try to get rid of such food items by artificial means like laxatives as these cause more problems as earlier intimated. Increased physical activity in general and specific exercise in last chapter of this book in particular could well be the answer in addition to other measures mentioned in this book.

We can liken constipation to a journey in a car. Normally before we embark on this journey, its imperative to make sure that the car is well serviced. We do not start driving without an MOT, adequate engine oil, brake fluid, well inflated tyres, functioning wipers, lighting, mirrors, coolant, seat belt and above all enough petrol in the tank or provision on the route for a top up should the need arise. These measures are within our control.

Unfortunately, the condition of the road is not entirely within our control. We can, of course, put some pressure on the government/ council to ensure that the roads are well maintained; with no pot holes, well displayed road signs, street lighting, and reduction of congestion to the barest minimum by providing alternative routes and increasing the number of lanes. Disobeying traffic regulations by overspeeding, unnecessary over taking or driving too slowly or lack of consideration for other road users are all within our ability to control. We can not blame anyone for any misdemeanour except, of course, other inconsiderate road users. The sudden appearance of fog, heavy rain or snow are not within our control but these are extremely rare as usually there is a weather forecast which warns about these conditions before we embark on our journey. We can therefore decide to postpone the journey or make provisions to adequately manage such adverse weather conditions.

In the same manner, it is within our control to eat a diet that is rich in fruits, vegetables, fibre and take adequate amount of fluids. We must also ensure that we do not live a sedantory lifestyle. Exercise, if well maintained is the engine of the body. It does not only improve our digestive system, but our cardiovascular (CVS), blood pressure, helps prevent heart attacks, strokes, cancer, obesity and diabetes. There is little doubt that regular exercise plays a major role in the fight against **constipation**. We can occasionally ditch the car and walk or ride a bicycle to work or the corner shop. These are other ways we can exercise.

Alternatively, we can join the Gym, or buy and use exercise equipment at home such as the exercise bike, skipping rope, treadmills, weight benches, weight equipment, barbell and dumbbell kits, rowing machines, elliptical trainers etc. "Variety is the spice of life" but in this case it may not be. Apart from the cost, it is better to concentrate on those equipment we think we can afford and use effectively than buy several equipment which we rarely use and which also take up much needed space at home.

Many aspects of road (colon) maintenance are within our ability to control. We should not hesitate to consult our doctor/GP if there is **a change in bowel habit.** This may be quite innocent but may also be due to tumours, polyp, diverticular disease, irritable bowel syndrome, obstructed haernia, Crohn's disease, ulcerative colitis etc. Just as the road surface can be managed and rehabilitated so are these conditions which may be related to constipation.

The next chapter deals with most of these conditions which can slow down the colonic traffic or cause a change in bowel habit Some are quite easy for the layman to spot especially after reading through this book. Others are more difficult and complicated and a doctor should always be consulted.

Most cases of constipation are due to benign causes which can easily be addressed by the methods in this book.

CHAPTER THREE

Causes of Constipation Due to Neoplastic Conditions in the Wall of the Colon or Rectum

Most causes of constipation are due to the slowing down of contents of the colon, occasioned by poor dietary contents of 3FEV and consequent excessive water absorption as well as lack of physical activity already discussed in chapter one.

This chapter deals with the causes of constipation due to neoplastic conditions (new growth- benign and malignant tumours) in the colon.

Some of the conditions are quite rare but after reading this chapter, most people should be able to know when there is **a change in bowel habit** especially constipation alternating with diarrhoea or obstinate constipation (very severe constipation) in which there is no passing of gas or faeces for several days **(obstipation)** or the passage of blood or mucus with or without faeces and may be accompanied with abdominal pain. These are some of the conditions when a doctor should be consulted.

We can now talk about some of the diseased conditions that can cause constipation.

It must be stressed here, however, that the polyps themselves are unlikely to be obstructive enough to cause constipation per se but could be the prelude to the malignant tumours that cause constipation or co-exist with them.

1. Diseased conditions of the colon and rectum. These are conditions in the large intestine that obstruct its contents and present with constipation. There are usually some other symptoms but sometimes these may be quite subtle.

(a) Tumours of the colon and rectum may be benign or malignant.

Benign tumours such as **polyps** should be viewed with suspicion as they may be pre malignant A polyp is a swelling arising from the mucosa of the intestine and has a stalk (pedicle) or described as pedunculated

Polyps may be classified as follows: -

Pseudopolyps or inflammatory polyps which are usually found in patients with diverticulitis, chronic dysentery, Crohn's disease and ulcerative colitis. As the name pseudo polyp implies these are not strictly true polyps. Notably the associated conditions themselves give rise to constipation via other mechanisms; see later

Hamartomatous polyps are either **Peutz-Jeghers syndrome** which is a familial condition or **juvenile polyps.**

Peutz-Jeghers Syndrome They are associated with pigmented lesions on the tongue, mucous membrane of the mouth or face. They are small and pre malignant and cancer can start even in young people and so it's important to see a specialist when there is a suspicion or a diagnosis has been made. They occur more often in the small intestine than the stomach or colon. The other group of hamartomatous polyps are juvenile polyps.

Juvenile polyps are also familial and occur early in life, more often in the male child than the female and tend to present in the first decade, usually in the lower sigmoid colon or rectum. Juvenile polyps occur in infants and young children and tend to be multiple. They are not thought to have any significant malignant potential.

Metaplastic polyps are generally small and are elongations of the mucosal glands and occur in any age either in the colon or rectum and are benign.

The most important group of polyps are the **adenoma and villous papillomas.** They are pre malignant and account for most cases of cancer of the colon and rectum. Adenomatous polyps are quite variable in size and small lesions tend to assume the colour of the mucosa of the colon but larger ones are darker and more vascular and tend to develop a stalk (pedunculated) as they enlarge. Villous papillomas are usually described as sessile, lacking any stalk or peduncle but more like a carpet. Lesions around the rectum tend to have greater tendency to malignancy and more so in males than females. The villous papillomas are usually single whilst the adenomas may be multiple. These lesions may actually coexist with cancer. They may affect any age group but more commonly first diagnosed in the sixth decade of life. Villous papilloma has a slight female preponderance.

I have to emphasise again the fact that although I have considered polyps under tumours capable of causing constipation: in actual fact this is unlikely unless they are very large or malignancy has supervened. They are more likely to be diagnosed during a routine investigation especially to find out the source of occult blood during a screening exercise. They may also present as blood in stool or **change in bowel habit**. Villous papillomas may present with a copious mucous discharge which may manifest as diarrhoea resulting in dehydration and electrolyte depletion especially potassium and to a lesser extent sodium and chloride. In large papillomas, dehydration and potassium depletion (hypokalaemia) may result in metabolic acidosis and severe ill health, mental confusion, weakness of muscles, lethargy and rarely kidney failure.

Familial Polyposis Coli-This is a hereditary condition in which multiple adenomatous polyps develop usually in the colon. The polyps are usually multiple and mostly pedunculated though sessile (having no stalk) lesions can also occur. As the name implies this condition runs in families with a dominant gene inheritance. It affects both sexes equally and transmitted by the sufferer to half the children. There are no polyps at birth but starts at about puberty and malignant change which is virtually inevitable occurs about the third decade.

A small proportion of suffers, about a tenth, have associated markers called Gardner's syndrome characterised by connective tissue tumours, bony exostosis, dermoid and sebaceous cysts which may actually occur before the polyps develop and may be the first warning signs.

As described above, initially, there are no obstructive symptoms and so constipation is not a feature and may only occur with the development of malignancy. **A change in bowel habit** with frequency, loose stool, accompanied with blood and mucus are likely to be the presenting symptoms. As already advised, **a change in bowel habit** is an indication to consult your doctor.

The distribution of colonic polyps virtually coincides with that of colonic cancer and reinforces the view that most polyps are pre malignant and are the precursors of cancer and therefore closely associated with it Most polyps about 80 percent occur in the recto- sigmoid region and about 70 percent of cancer occur in the same region. The descending colon has about 5 percent of polyps and cancer. However as we move towards the right, a slight disparity becomes apparent, with the transverse colon bearing a little more burden in cancer distribution compared to polyps with a ratio of about 2 to 1. This continues to the ascending colon with roughly 3 to 1 ratio.

Malignant Tumours (Cancer of the colon and rectum)

Cancer of the large intestine can present in a variety of ways. The presentation may be one of a **change in bowel habit** with constipation alternating with loose stool or diarrhoea. More often there may be non specific ill health including anaemia, weakness and weight loss. Sometimes there is abdominal distension, a mass, pain, flatulence (excessive wind) and vague abdominal pains or peritonitis arising from perforation. In some cases there is rectal bleeding or blood and mucus in stool.

The clinical presentation of colonic cancer depends on anatomical site or distribution, the type of cancer and the stage of the cancer at that point in time. The typical obstructing cancer of the colon with **constipation** and possibly abdominal distension may only be seen possibly in a tenth of cases.

As noted above, clinical presentation depends on the type of cancer and the position of the cancer in the colon or rectum. A brief summary will suffice as a full description is beyond the scope of this book.

There are 5 main types of cancer of the colon viz: -polypoidal, annular, diffuse, ulcerative and colloidal.

Polypoidal carcinoma This is the commonest type. At the outset, it resembles the polyps described in benign lesions.. It tends to infiltrate less than the other types.

Annular carcinoma-is circumferential. Typically because of the constriction across the circumference of the colon, it is more likely to present with absolute **constipation** (obstipation- see above) or symptoms of complete intestinal obstruction.

Diffuse carcinoma This affects mostly the muscle coat leaving the lining of the colon intact. It may be associated with chronic inflammatory bowel disease like ulcerative colitis.

Ulcerative carcinoma like the name implies, is like an ulcer but with a typical malignant characteristic with raised everted edges and infiltrating into the colonic wall and later extending into the lumen of the colon.

Colloidal carcinoma is a rare mucus secreting cancer and has a variety called mucoid carcinoma. Presentation is likely to be the secretion of excessive mucus before obstructive symptoms supervene.

Secondary Deposits of cancer from primary intra abdominal cancer-This may be a direct invasion of the sigmoid colon or caecum from infiltrating ovarian cancer leading to constipation or complete bowel obstruction.

I shall now look at presentation of cancer of the colon based on the part of the colon involved.

Cancer of the caecum is likely to be non obstructing and present with anaemia, weakness and weight loss. There may be some abdominal pain as a result of tumour infiltration or flatulence with dyspepsia. Occasionally there may be slight looseness of the stool.

In the transverse and left colon, the presentation is more likely to be one of abdominal pain, distension, or a **change in bowel habit** with mucus and blood. As the carcinoma in this part is more likely to be of the annular variety, presentation especially in the left colon may be one of absolute **constipation** if the tumour has encircled the whole circumference of the colonic wall.

Cancer of the sigmoid colon and rectum especially if annular may present with absolute **constipation** (obstipation) in the late stage but initially a partially obstructing tumour presents with **a change in bowel habit**-diarrhoea alternating with **constipation,** or blood and mucus in stool.

I wish to emphasise that these presentations are the typical or characteristic and mostly in the early and not so early stages of the cancer In neglected and late stages, any of the various types of cancer in any part of the colon and rectum are still capable of presenting with **constipation or obstruction** if the cancer has advanced to a stage where colonic contents can either not pass through a badly diseased and obstructed colon or the delay in the journey has culminated in excessive absorption of water resulting in very hard stool which further impedes its passage. Analogically this is like friction or resistance slowing down the movement of machinery or humps and bumps or uneven road riddled with pot holes slowing down the speed of a car.

At this stage I like to mention another important factor capable of causing, contributing or worsening **constipation** in cancer patients. Apart from obvious anatomical obstruction presented by cancer itself, chemotherapy and pain relieving medication especially the opioids (narcotics)-morphine and codeine based analgesia present further problems of constipation. I shall deal with this further under medications that cause constipation. Meanwhile in the next chapter I shall deal with benign conditions of colon and rectum that can cause constipation.

CHAPTER FOUR

Benign Conditions of the Colon and Rectum that May Cause Constipation

In this chapter I shall deal with benign conditions (not tumours) of the colon and rectum that can present with constipation. You would note, however, that I included benign tumours (essentially polys) in previous chapter for convenience. This chapter therefore deals with other benign conditions of the colon and rectum but excluding polyps (tumours)

It is to be emphasised that all these conditions and those in the previous chapter are rare causes of constipation and are included for completeness of this book. They may be of interest to doctors and research scientists. The lay readers could glance over chapters 2-7 as they may still be useful as an "exclusion exercise" so that by the time they see their GP/doctor, they have good ideas of what to talk about and what the causes of their constipation are likely to be. In the vast majority of cases, perhaps up to 90 percent, the cause of your constipation is unlikely to include the contents of these chapters.

Conversely chapters one and eight to twelve of this book are to be "carefully, digested, absorbed, assimilated and utilised" They form the building blocks on which this book is based for the layman/ woman for which it is intended.

There are a few benign conditions of the colon that cause intestinal obstruction and if left untreated may present as constipation. Some of these conditions present acutely as intestinal obstruction and others insidiously as constipation. I shall mention both categories and briefly explain possible causation and presentation. These are: -

(a) Diverticular disease of the colon
(b) volvulus of the sigmoid colon
(c) Intussusception
(d) scarring
(e) stricture
(f) Megacolon and mega rectum which may be acute (as in toxic mega colon) or chronic
(g) Hirschsprung's Disease-congenital mega colon
(h) Granulomatous colitis
(i) ulcerative colitis
(j) Severe Dysentery - leading to electrolyte deficiency and colonic pseudo obstruction
(k) colonic ischemia- more common in the elderly
(l) Chagas' Disease

(m) Crohn's disease

(n) Rectal Inertia

(o) Anal Fissure

(p) Endometriosis- implant of uterine tissue into sigmoid colon or rectum leading to narrowing of colonic lumen causing **constipation** or obstruction.

Diverticular Disease of the colon are herniations of colonic mucosa through circular muscles at weak points where blood vessels enter colonic wall.

The sigmoid colon is affected in about 90 percent of cases but the rectum which has a complete muscle wall is never affected. Diverticular disease is possibly due to disordered colic motility which is facilitated by refined diet especially flour and sugar typical of European diet.

Diverticular disease is rare in native Africans where there is high dietary fibre intake. Studies have shown that native Africans have about 3 times the European daily stool weight and transit time of about 35 hours compared to European of 3-6 days. This accounts for the low incidence of diverticular disease in rural Africans. These differences are not thought to be racial because the disparity disappears on adopting European diet. The differences may also account for the lower incidence of reported colonic cancer in Africans due to shorter transit time and therefore less contact and absorption of any carcinogens in food.

Diverticular disease of the colon increases with age; it is rare before the age of 30. By 40 years of age about 5 percent of Europeans have this condition and is maximal after 60 years of age.

Almost 90 percent of individual are asymptomatic but in the remainder several presentations occur; there may be a feeling of distension which is relieved by flatus, pain, small hard stool, diarrhoea and mucus alternating with **constipation**.

When this disease becomes complicated by inflammation, diverticulitis supervenes and is marked by severe pain in the left iliac fossa, but the pain could be dull and occur centrally or even on the right lower abdomen.

There is usually low grade fever, occasionally tenderness and rigidity as well as a left sided mass analogous to appendicitis and aptly referred to as "left sided appendicitis"

Volvulus results from a rotation of the intestine around its axis. It may occur in the very young **(volvulus neonatorum)** or the **adult**. In the neonate the caecum and small intestine are involved. It is acute and could be catastrophic. As this is an **intestinal obstruction** per se, I shall deal a bit more with the adult form which could be acute (more often) and chronic (less often). In the latter case, the presentation is one of **constipation.**

The adult volvulus can occur in the small intestine, usually the ileum but I shall concentrate on the more common volvulus of the colon -**volvulus of the caecum and of the sigmoid colon.**

Volvulus of the caecum occurs when there is laxity and excessive mobility of the caecum and ascending colon resulting in a twisting of the large intestine, often in a clockwise direction which obstructs the ascending colon and if a second twist occurs, the ileum is obstructed. Age of presentation varies widely from adolescent to about 90 years of age. It is about twice as common in the female. It is usually acute with abdominal pain, nausea and vomiting but many cases also present chronically with **constipation.**

In about a quarter of cases, there is a palpable mass, usually on the right iliac fossa but during the process of torsion the caecum may become very mobile and move centrally or to the left. The caecum become ballooned or distended and may perforate resulting in peritonitis.

Volvulus of the pelvic colon -**sigmoid volvulus** is similar in mechanism to volvulus of the caecum except that it is usually anticlockwise in direction, occurs more in middle age and the elderly and males are more often affected than females. It is less common in the West but seen more often in South America, Eastern Europe and Africa. Sigmoid volvulus may occur in bouts of acute attacks which are relieved by the passage of large amount of flatus and bowel motion and presumably a partial volvulus untwists. More often it is acute with severe abdominal pain, and abdominal distension occurring more often during straining at stool. Initially there is left sided abdominal distension which after a few hours becomes generalised. **Constipation** becomes absolute (obstipation)

Intussusception is another condition which is far more likely to present as acute intestinal obstruction than **constipation**. Intussusception is a surgical emergency when one portion of the gut invaginates into another portion immediately adjacent to it; usually the proximal (upper part) into the distal (lower part)

There may be an obvious cause such as a polyp, a sub mucous lipoma, a carcinoma or a Meckel's diverticulum.

This condition is far more likely to occur in childhood; usually an infant between 4 and 9 months of age. The male child is more usually affected than the female.

There is an outer tube (the intussuscipiens) that receives the inner and returning middle tube (the intussusceptum)

Intussusception is described according to the part of intestine involved as ileo-ileal, ileo-colic, ileo-ileo-colic and colic-colic. Initially a normal stool is passed but after several hours there is passage of blood mixed with mucus and described as "red currant jelly" stool which typifies this condition. It is quite a painful obstructive condition and the infant usually screams with pain.

If the condition is unrecognised and treated, after 24 hours there is abdominal distension and vomiting becomes more frequent. If still undiagnosed, there is **absolute constipation (obstipation),** gangrene, perforation of the gut and death.

Scarring of the intestine occurs when the normal healthy tissue is replaced by fibrous tissue which narrows the lumen of the gut in the affected area and is closely related to **stricture** which is a circumscribed narrowing or stenosis of the gut and may be congenital or acquired and the latter is usually due to infection, chronic inflammatory bowel disease, muscular spasm, mechanical or chemical irritation.

Chronic inflammatory bowel diseases which are usually responsible for these conditions in the colon, encompass a number of conditions including well defined conditions like ulcerative colitis, ulcerative proctocolitis and Crohn's disease and the ill- defined condition of **irritable bowel syndrome (IBS)**. The well defined chronic inflammatory bowel diseases are complicated by stricture and scarring leading to chronic **constipation.**

Irritable bowel syndrome has been shown by barium studies to be responsible for diffuse narrowing, increased segmentation or corrugation (wrinkling) of distal colon. These lead to narrowing of the colon and the passage of small hard faecal pellets-a feature of **constipation.**

Hirschsprung's disease (Primary or congenital mega colon) is a congenital condition which usually manifests in the first 3 days of life with gross dilatation and hypertrophy of the pelvic colon which may extend into the descending colon. The condition is more common in male children and is due to **complete absence of parasympathetic ganglion** cells usually in the rectum and anal canal which become constricted or spastic. Above the rectum there is an intermediate zone in which there are sparse parasympathetic ganglion cells (the intermediate zone) and above there is the dilated part of the pelvic and descending colon where the parasympathetic ganglion cell are present as normal.

This condition becomes evident when the baby fails to pass meconium (greenish discharge in the newborn baby and consists of bile, epithelial cells and mucus which is the first discharged before actual stool) in first 3 days of life. It is characterised by **absolute constipation** (of meconium) However, the insertion of a small tube or finger would stimulate the passage of a little meconium and small toothpaste- like motions after straining which is usually obvious. After this, there is gross abdominal distension and visible peristalsis (excessive gut movement) become evident due to intestinal obstruction. The condition should not be allowed to progress to this state without hospital consultation and admission as Hirschsprungs disease requires specialist management and surgical treatment. The surgical procedure is beyond the scope of this book.

Acquired Megacolon and Megarectum are secondary dilatations of the colon or rectum which may be acute or chronic and characterised by faecal impaction and subsequent **constipation** which may be absolute.

Acute or toxic mega colon are usually encountered due to the following conditions: -

(a) acute ulcerative colitis
(b) colonic ischemia- usually seen in the elderly.
(c) severe dysentery -leading to severe fluid and electrolyte (especially potassium) losses and causing colonic pseudo obstruction
This may also be encountered in severe fluid and electrolyte loss due to excessive use of non potassium sparing diuretics.
(d) It is also a feature of anti cholinergic drugs overuse
(e) Granulomatous colitis.

Ulcerative Colitis-Let me give a little more space to this condition which is not uncommon. This disease which can start acutely often ends up as chronic-**chronic inflammatory bowel disease**. It is characterised by minute ulcers in the colon which may coalesce to form larger ulcers. With time it extends into the sub mucosa of the colon and causes muscle spasm leading to the formation of strictures and subsequent **constipation.**

Most cases start in the rectum and spread proximally to the caecum and if the ileocaecal valve is incompetent the ileum becomes involved resulting in ileitis (inflammation of the last part of the small intestine-the ileum).

The cause is unknown but there appears to be a genetic predisposition as it tends to be familial and virtually confined to Caucasians. Females are more often affected than men and the onset is usually between 20 to 40 years of age. Few cases commence in childhood.

Acute ulcerative colitis is somewhat rare (5 percent) and starts with a high temperature of about 39 degrees Centigrade. There is severe diarrhoea with blood, pus and mucus. Severe ill health is the norm. This acute type may be responsible for **toxic dilatation of the colon** and **constipation** which supervenes.

Chronic ulcerative colitis is more common (95 per cent of cases) The initial attacks are mild to moderate but exacerbations take its toll between intervals. A change in bowel habit ensues especially diarrhoea which may be up to 20 times a day with tenesmus. The patient becomes very ill with dehydration and emaciation and the degree of ill heath being proportional to the length of bowel involvement.

It is characterised by lesions between healthy bowel but as it progresses virtually the whole colon becomes involed. There are narrowing and contraction of some parts of the colon (pipe stem) altered mucosal outline and pseudopolyposis (false polyps). These features are better demonstrated radiologically. The narrowing engendered by these chronic features are responsible for the **constipation** at this stage of the disease. This should be contrasted with the **constipation of toxic mega colon encountered in the acute stage.**

The development of fibrous stricture as a complication of chronic ulcerative colitis is yet again an additional pathology in the evolution of **constipation.** Additional and more serious complication but fortunately not very common is epithelial dysplasia (pre malignant change in the lining of the colon) and the development of **carcinoma** which increases with the duration of disease and the extent of involvement of the colon. The over all risk is less than 4 percent but after 20 years this rises to about 12 percent. Carcinoma is an additional factor in the manifestation of **constipation.** Because this book is intended mostly for the **lay reader** and bearing in mind the mention of carcinoma as a long term complication, it is pertinent to emphasise the fact that the surgical procedure of a total recto- colectomy (removal of the rectum and colon) with ileostomy virtually obviates the development of a carcinoma if carried out before chronicity sets in - perhaps 10-20 years from onset especially if there is any suggestion of epithelial dysplasia. Therefore people who suffer from this condition should not regard it as a life sentence.

<u>**Chagas' Disease**</u>-Is caused by a protozoal flagellate -Trypanosoma Cruzi which gets into the blood stream through an insect vector (aptly called assassin bugs and kissing bugs) which bites its victims during the night. From the blood stream the organism proceeds to the heart muscle (myocardium) and smooth muscles of the intestine. It then proliferates several times and up to 10 percent of its victims die at the acute phase of the infestation. (hence assassin bugs). The chronic disease causes dilatation of the smooth musles especially the gullet (oesophagus) and colon. Infestation of the heart muscle could cause congestive heart failure and heart block.

Infestation of gut muscles causes mega oesophagus and mega colon (dilatation of these parts of the intestine) by severe reduction in the number of myenteric ganglion cells. The reduction in gangion cells leads to hypertrophy and stagnation of contents (compare Hirschsprung's Disease) In the colon this stagnation causes **chronic constipation** and distension of the abdomen. Unless prompt diagnosis and treatment are instituted, the patient becomes very ill. Realistically, prevention remains the best form of Public Health measure by eradication of the insect vector. This disease is prevalent in South America, especially Agentina, Brazil and Chile.

Crohn's Disease- Like ulcerative colitis above, Crohn's Disease is a chronic inflammatory bowel disease displaying most of the features of ulcerative colitis; namely acute fulminating (granulomatous) colitis with or without toxic mega colon as well as, narrowing, fibrosis, stenosis (so called Kantor's string sign in barium enema especially of the terminal ilium) and stricture. These features are responsible for **constipation** which may occur in Crohn's disease. It must be emphasised however, that only about 5-10 percent of cases would have **constipation** as a significant feature.

Typically this condition, presents with pain usually in the right iliac fossa which closely mimics appendicitis as most cases tend to start in the terminal ileum which is anatomically in the same area as the appendix. Diarrhoea occurs in 70-80 percent of cases whereas in appendicitis diarrhoea is rare though an important feature in pelvic and post ileal appendicitis.

Unlike ulcerative colitis, Crohn's disease is transmural (penetrates deeply into the wall of the colon). Also nearly every part of the gastrointestinal tract can be affected whilst ulcerative colitis is a mucosal disease (affects the outer wall-the mucosa only) confined mostly to the colon and rectum. The degree of invalidity is similar to ulcerative colitis.

It is more prevalent in Europe and the USA, where the incidence is increasing. It is more common in the white races and Jews than blacks and non Jews.

Crohn's disease is characterised by "skip lesions" where there are diseased areas of the bowel with intervening normal areas. Mucosal oedema gives it a "cobblestone appearance". In addition to the oedema, there is sloughing and ulceration of the mucosa. As a result of the transmural nature of this disease the ulcers penetrate deeply into the muscle layers and referred to as "fissure ulcers" and account for the tendency to perforation and fistulation into adjacent viscus with different types of fistula formation-perianal (around the anus) enterovesical (part of the gut into the bladder), enteroenteric (one part of the gut to another), enterovaginal (gut to the vagina), entero cutaneous (gut to outside skin).

Perianal anal complications; namely: - perianal abscesses, perianal tag, anal fissures, internal haemorrhoids, and fistulae-in -ano are responsible for many cases of **constipation** as these patients avoid opening their bowel due to pain and discomfort.

Rectal Inertia is characterised by **chronic constipation** in which there is overloading of the rectum, sigmoid and descending colon and occasionally the right side of the colon could be involved depending on the degree or seriousness. Children are mostly affected but unlike Hirschprung's disease, it does not manifest in the first few days of life: the children are usually older and look remarkably well despite their bowel problem. Some cases are found in adults, some of who may have psychological or mental health problems or are epileptic.

Some of the children may have had inadequate or over zealous toilet training despite unrecognised peri-anal condition such as painful fissure-in-ano or congenital anal stenosis which make normal bowel movement difficult or painful.

On rectal examination there is a large faecal mass mostly in the recto sigmoid colon and perianal soiling is common but abdominal distension is either mild or hardly apparent. As a result of incomplete evacuation, there is a build up of faeces in the rectum and overt incontinence ensues.

Manual evacuation is usually required initially followed by saline rectal washouts. It is important to exclude any perianal conditions or any other triggers and manage effectively if present

I have already mentioned **anal fissures (fissure-in -ano)** Briefly this is a small tear or an ulcer in the long axis of the lower anal canal. They are usually in the midline posteriorly or anteriorly. They may sometimes be caused by **constipation** in which the passage of a hard faecal mass causes a tear of the anal mucosa and shortly after, a vicious cycle develops in which the pain caused by the fissure stops the individual from emptying his/her bowel leading to more and often worse **constipation.**

In women, **consipation** can also be caused by <u>**endometriosis.**</u> **Endometriosis** is a condition when endometrial tissue (lining of the womb) is found in ectopic sites.

When endometriosis occurs in the bowel wall, it causes narrowing of the lumen of the gut and stricture develops and results in **constipation.**

There are several theories to explain the mechanism of development of endometriosis but that will not add to the understanding of its effect on the colon resulting in constipation which is what concerns us. The theories of causation is therefore beyond the scope of this book.

Endometriosis may also be found in other sites particularly the ovary which is the commonest site. It may also be found in pelvic peritoneum especially the the pouch of Douglas, uterosacral ligament, the bladder and could be a cause of blood in urine (haematuria). Other sites include the vagina, abdominal wall scars, perineum and the uterus itself where it can be found in the muscle of the uterus (myometrium).

Normal endometrium undergoes cyclical changes and bleeding occurs during menstruation when shedding of the endometrial tissue occurs. Endometriosis follows the same cycle. Cyclical bleeding from the bowel occurring at the same time as menstruation is therefore a feature of endometriosis. This also applies to other sites where endometriosis occur. Presentation may be painful menstruation. (dysmenorrhoea), infertility, painful coitus (deep dyspareunia) or excessive menstrual flow (menorrhagia).

Usually endometriosis tends to affect women in their 30s and 40s but may affect even teenagers. It does not occur before puberty. Caucasians are usually affected and is rare in non white races.

Endometriosis is oestrogen dependent and regresses during pregnancy due to regression of endometric tissue and an increase in progesterone. For similar reasons, it tends to regress after the menopause.

As already discussed, the above conditions are not the usual causes of "every day" constipation encountered in clinical practice but they do occur and present particular problems due to lack of recognition and anticipation. The only way of overcoming the problems of diagnosis is to keep these conditions in mind by a good history, clinical presentation and investigations if necessary.

LARGE INTESTINE

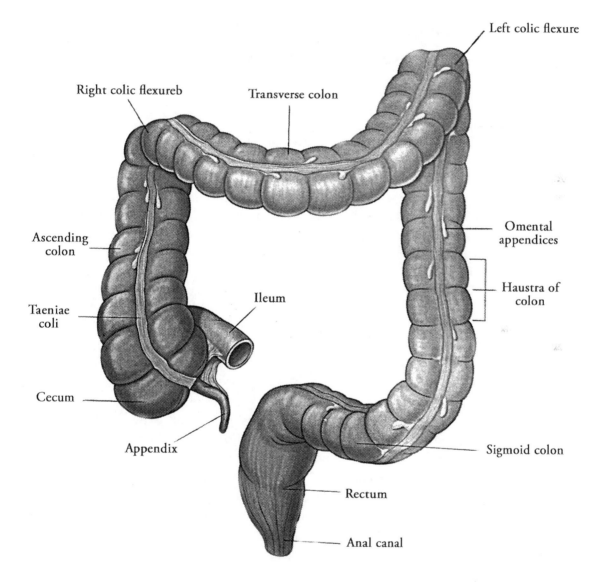

Left colic flexure

Right colic flexureb

Transverse colon

Ascending colon

Omental appendices

Haustra of colon

Taeniae coli

Ileum

Cecum

Appendix

Sigmoid colon

Rectum

Anal canal

With Kind Permission of ELSEVIER Churchill Livingstone.

CHAPTER FIVE

Neurological, Endocrine, Metabolic and Systemic Diseases that Cause Constipation

There are a number of systemic diseases and general conditions that cause constipation. These diseases either affect the whole body or specific systems of the body that are not directly related to the colon or rectum but capable of influencing them directly and often indirectly.

These diseases and conditions cause **constipation** by slowing down colonic motility, the movement of contents of the colon or rectum and therefore as seen in chapter one lead to increased absorption of water and passage of hard faeces. Three separate processes are involved here, which all cause constipation.

They can be divided into the following groups: -

All three conditions and diseases cause the slowing down of contents of the colon, rectum and anus.

(a) Neurological Diseases
(b) Endocrine and metabolic diseases
(c) Systemic Diseases

It is important to understand that colonic motility is controlled by neuro -hormonal mechanism which is involved in colonic response to the entry of food into the gastrointestinal tract, the **"gastro -colic reflex".** There is increased emptying of ileal contents into the caecum, accompanied by increased **mass movement** and the urge to defaecate. **Gastro colic reflex** is dealt with in more details in chapter twelve. Please refer to the 3 types of motor activity of the colon described earlier -Segmentation, mass movement and retrograde peristalsis (see chapter two) In terms of effective movement, mass movement is probably the most important because it moves contents of the transverse and descending colons distally in a strong and ring-like contraction and probably occurs 1-3 times a day in response to food. This is the de facto peristalsis.

Neuro-hormonal mechanism is not the only determinant of colonic motility: faecal bulk, physical activity and emotional state also play their parts.

(a) Neurological Diseases that cause constipation

Stroke
Spinal Cord Injuries

Multiple Sclerosis
Parkinson's Disease
Pseudo Obstructive Syndrome
Spina Bifida- in children
Cauda Equida Lesions

Inorder to understand how (a) **Neurological diseases** cause **constipation,** it is important to appreciate that the colon, rectum and the anus like other parts of the gut are richly supplied by nerves and mass movement and other movements of the colon would be greatly impaired or impossible when there is a diseased process or interruption to the nerve supply of the colon, rectum and anus.

Briefly, there are 2 main types of autonomic nerve supply- the **sympathetic** and **parasympathetic.** They constitute the autonomic nervous system. These 2 systems maintain a balance between doing too much or too little and being autonomic their activities are outside our control. They are in opposition or complementary to each other. With exceptions, generally the sympathetic system "excites" and carries out its actions very quickly whilst the parasympathetic system acts with greater caution, applies the brakes as necessary and dampens the activities of the organs like the heart, lungs, eye etc. This generalisation is in fact not strictly true: in the gastrointestinal tract, the opposite is the case.

The **sympathetic** nervous system inhibits peristalsis and constricts internal sphincters. Inhibition of peristalsis and constriction of internal sphincters are actions which would promote **constipation**

The **parasympathetic** system on the other hand has a calming effect and a return to regular function. In the upper gastrointestinal tract (GIT), it enhances digestion, dilates blood vessel thereby enhancing blood flow. In all parts of the GIT, it increases motility (peristalsis) and relaxes sphincters thereby returning to normal function. However, if the parasympathetic nerve activity was increased beyond normal activities it would tend to promote diarrhoea. Please note why there is constipation in Hirschsprung's disease in chapter four where there is congenital absence of parasympathetic ganglion cells in the rectum and anal canal where the "constipating action" of the sympathetic nerve is unopposed.

Generally the two systems work hand in hand and maintain a balance of normal bowel activity. In normal subjects, they cooperate rather than antagonise each other and thereby obviate the dominance of one system over the other. In diseased states as would be seen later this balance is shifted to one side or the other leading to either diarrhoea or constipation. We shall concentrate more on **constipation** which is the subject of this text.

The abdominal viscera (the gut and associated organs) have both extrinsic (outside the wall of the gut) and intrinsic (within the wall of the gut) nerves

The **extrinsic innervation (**nerves outside the wall of the gut) of the colon involves receiving motor impulses from and sending sensory information to the CNS (Central nervous system- brain and spinal cord) The motor impulses are concerned with motor activities like peristalsis and the churning of colonic contents whilst the sensory component is concerned with chemical and mechanical conditions in the colon.

The intrinsic innervation of the colon are nerves embedded in the wall of the gut and are part of the division of autonomic nervous system (sympathetic and parasympathetic nerves). They regulate digestive tract activities by self sufficient network of sensory and motor neurons. They are therefore called the intrinsic or enteric nervous system.

Generally the visceral nerves send sensory information back to the CNS through **visceral afferent fibres** and receive back motor impulses from the CNS via the **visceral efferent fibres**

The visceral efferent fibres are part of the sympathetic and parasympathetic parts of the autonomic division of the peripheral nervous system

I shall now discuss briefly what nerves constitute this system, their anatomical locations and functions and possible consequences of interruption or diseased states

The colon and rectum are supplied by both the sympathetic and parasympathetic nerves.

The caecum and ascending colon receive **sympathetic nerve** fibres from the dorsal ganglia of the last six thoracic nerves -T7-T12.

The **parasympathetic fibres** are derived from the coeliac branch of the posterior vagus nerve

The left colon and rectum derive their nerve supply from the proximal 3 lumbar ganglia - L1-L3. There are also contributions from the pre aortic, inferior mesenteric and pre sacral nerves.

The plexuses themselves derive some branches from the sacral parasympathetic nerves.

The anal canal derives the motor supply of its internal sphincter from the pre sacral plexus whilst the external sphincter receives its supply from the inferior haemorrhoidal branch of the internal pudenda nerves and perineal branch of the 4th sacral nerve (S4). This is also the supply to the levator ani muscle. Sensation below the anal canal is via the haemorrhoidal nerve whist the autonomic nervous system supplies the area above this.

I shall now briefly discuss how the above conditions cause constipation and possible relationship to colonic innervation or any other associated factors.

Stroke is a serious neurological disease and results from ischemic infarction (loss of blood supply) or bleeding into a part of the brain. The common manifestation of stroke is due to a deficit in brain functions which may manifest as hemiplegia (weakness of one side of the body usually leg and arm), aphasia (loss of speech) or loss of sensation on one side of the body etc.. It is obvious that the impairment of the CNS in stroke would affect virtually all neurological connections of the colon to the brain particularly the visceral afferent and efferent fibres noted above which are important in peristalsis. The autonomic (parasympathetic) nervous supply to the colon would also be impaired resulting in **constipation**.

Spinal cord injuries usually involve some interruption in the same fibres already noted in the case of stroke and the over all effect is the same and results in **constipation**, again due to its effect on slowing down of peristalsis, and consequently movement of bowel contents, leading to excessive water absorption of colonic contents.

In **Multiple Sclerosis (MS)** there is an autoimmune attack of myelin producing brain cells leading to patches of sclerosis or plagues of the CNS. It is one of the commonest neurological causes of long term disability. It is thought to have a genetic and environmental basis being rare in the tropics and common in temperate climates. Presentation depends on the area of CNS affected and the size of the sclerosis. Thus the clinical manifestations may include visual impairment or loss, diplopia (double vision), nystagmus, weakness of limbs, paresthesia (abnormal sensation like burning, tingling, pricking or tickling) and mood changes etc

It is obvious from the above description of MS which is essentially a disease of the central nervous system with widespread neurological effects that the bowel would also be affected in the same way as in stroke or spinal cord injuries. Thus **constipation** is likely.

Parkinson's Disease. This is a degenerative disease of the basal ganglia (part of the brain) It is characterised by tremor, rigidity (increased tone) and bradykinesia (slow movements) This classical syndrome may be absent initially but preceded by tiredness, depression, slowness of mental activities, and small handwriting (micrography). The tremor is usually initially one sided affecting the hand and later the leg, mouth and tongue. There is a characteristic difficulty in rapid fine movements which may manifest as difficulties in shaving, writing, undoing or fastening buttons as well as slowness in gait.

The rigidity causes stiffness and the posture becomes flexed. In advanced disease, speech becomes indistinct and soft and there may be some difficulty in understanding their speech.

As noted above, since Parkinson's disease affects the brain, peristalsis could be affected resulting in prolonged journey of colonic contents and consequent **constipation**.

Intestinal Pseudo obstruction. This is a condition of uncertain aetiology (cause) is characterised by defective ability of contents of the intestine to go through the bowel. It may present with signs and symptoms of intestinal obstruction but without any lesion in the lumen of the intestine. It may start at any age with nausea, abdominal pain, distension, vomiting, dysphagia (difficulty in swallowing) diarrhoea and **constipation.** The clinical picture depends on the part of the gut affected. It is a rare condition but mentioned here for completion.

Spina Bifida. This is a condition in the newborn child where there is failure of the fusion of vertebral (spinal) arches and is frequently associated with poorly formed or mal-developed spinal cord and membranes that surround the spinal cord. Depending on the degree of this mal development, many of these babies do not survive. Mal development of the spinal cord results in either defective peristalsis and/or sphincter problems. We have already alluded to the role of the central nervous System (brain and Spinal Cord) in the normal functioning of the colon especially peristalsis. Spina Bifida either results in diarrhoea or **constipation** or both.

Cauda Equina Lesions. Below the end of the spinal cord, the anterior and posterior roots of the lumbar, sacral, and coccygeal nerves pass downwards to emerge at their exit in the spinal canal. The cluster of these roots of nerves is the cauda equina. As noted above, parasympathetic innervation increases motility and relaxes sphincters. Interruption of sacral nerves which run in the cauda equina and form part of the parasympathetic supply to the rectum and anal canal would result in **constipation** due to poor colonic/ rectal motility and tightening of the sphincters.

(b) **Endocrine and Metabolic Diseases**

There are a number of endocrine (ductless gland) and metabolic diseases that cause constipation as a result of mechanisms which are different from the above. Some of these mechanisms are ill defined but generally due to the general malaise and ill health that are the hallmarks of these conditions. In some cases the manner in which they cause **constipation** can generally be elucidated. This next section is aimed at unravelling some of the ways which the causative mechanisms are brought about. These conditions are listed below.-

Diabetes
High blood calcium level (hypercalcaemia)
Kidney Failure
Under active Thyroid (Hypothyroidism)
Inability to control Diabetes by diet or medication

Diabetes mellitus is a condition in which there is lack of or inadequate secretion of insulin, diminished utilization or development of resistance to insulin despite normal endogenous levels.
There are 2 types - types 1 & 2
Type 1 diabetes (formerly called insulin dependent diabetes mellitus-IDDM) is usually immune mediated or of unknown cause (idiopathic).
It is associated with profound deficiency of insulin and therefore requires replacement with insulin (injection).
No age is exempt.
In type 2 diabetes (formerly called non -insulin dependent diabetes mellitus -NIDDM), individuals retain ability to secrete insulin but there is insulin resistance or impaired sensitivity. It can therefore be treated without insulin injection by exercise, diet, and sometimes tablets. In the long term about 20 percent of type 2 diabetics will develop insulin deficiency and thus require insulin injection.
The common symptoms include frequency, excessive urination (polyuria), excessive thirst (polydipsia), unexplained weight loss in some patients, genital infections (especially thrush) and lethargy. These symptoms should alert you to see your family doctor.
Diabetes is associated with increased calories intake and lack of physical activity (exercise).
Diabetes causes **constipation** mainly because of the complication of **autonomic neuropathy** which is related to poor diabetic control. There is also the slowing down of both motor and sensory conduction. In the long term the central nervous system (CNS) is affected. The lower motor neurone of the lumbosacral plexus is affected by acute infarction in complicated cases
As explained above the effect on the nerve supply to the gastrointestinal tract (particularly the colon) including the autonomic nervous system could cause **constipation.**
In some patients there is general debility or the development of neuropathic cachexia in which the patient is very ill and hardly able to get out of bed. As noted before, physical inactivity or lack of exercise is a potent cause of **constipation.**
I have spent a little more time on diabetes because of its increasing incidence and prevalence due to life style factors especially sedentary life style or lack of physical activity. Fortunately there is also increasing awareness now, and diagnosis.

I believe that it is also important to emphasise the early recognition and see the family doctor for diagnosis, fast and early management which could be for a life time.

High serum (blood) calcium (hypercalcaemia) which may present with polyuria (excessive urination), polydipsia (excessive thirst) are similar to symptoms of diabetes above.

There may also be renal colic (kidney stones), lethargy, nausea, dyspepsia, anorexia, peptic ulcer disease, depression and **constipation**. Constipation could be due to a number of factors but particularly to dehydration occasioned by polyuria. Dehydration is likely to lead to dry hard faeces. See the "H" of SHIID in chapter one.

For the interest of medically minded readers I shall briefly mention the causes of hypercalcaemia.

Behind the thyroid gland lie 4 small glands-the parathyroid glands which help to regulate serum calcium levels. Enlargement of these glands in primary hyperparathyroidism (usually adenoma, hyperplasia, occasionally carcinoma) or tertiary hyperparathyroidism is one cause of hypercalcaemia. The clinical features of hyperparathyroidism are usually described as "bones, stones, abdominal groans and psychic moans".

Other causes include secondary cancer in bone (from bronchus, kidney, breast, thyroid and prostate.

Hypercalcaemia is also a feature of vitamin D intoxication, myelomatosis, sarcoidosis, thyrotoxicosis, drug induced (ion exchange resin, thiazides, Lithium), Paget's disease of bone, ectopic production of parathyroid-like substance by non endocrine tumours, Addison's disease (adreno-cortical hormone insufficiency) and familial.

Kidney Failure. This may be acute or chronic. In acute renal failure there is deterioration in renal function over several hours or days resulting in low urine volume (oliguria) usually less than 400ml / 24 hours. Biochemically there is a rise in plasma urea and creatinine.

Acute renal failure may be pre-renal, renal or post renal. Pre- renal acute renal failure is common and usually due to a number of factors especially circulatory collapse as in severe blood loss, sepsis (infection), severe illness, trauma, cardiac failure, liver cirrhosis or surgery. Under such circumstances, the body tries to maintain normal blood volume by absorption of water back into the circulation and in this case from the colon also, resulting in dry hard stool and **constipation.** Also diet and fluid control in some patients with renal failure may impact negatively on their intake of fruits, fibre and fluids (3F above) and cause **constipation.** In some patients with renal dialysis, the presence of fluid in the peritoneum may cause peristalsis to slow down increasing the likelihood of **constipation.**

Renal causes are usually due to acute tubular necrosis which is a damage to renal tubular cells which may be due to low blood volume or specifically hypo perfusion of the kidneys and nephrotoxins especially due to drugs (amphotericin, tetracyclines, amino glycosides, contrast agents, also uric acid crystals, haemoglobinuria following rhabdomyosis (muscle death) Other causes include malignant hypertension, myelomatosis, haemolytic uraemic syndrome, glomerulonephritis, thrombocytopenic purpura, interstitial nephritis and hepatorenal syndrome.

The post renal cause of acute renal failure is due to urinary tract obstruction which may be due to renal stones especially in a single functioning kidney, enlarged prostate and post surgery.

In chronic renal failure where there may be exacerbation of the above factors, there is also general ill health with poor nutrition and fluid intake leading to **constipation.**

Hypothyroidism. This is an under active thyroid (an endocrine gland in front of the neck)

Hypothyroidism may present with symptoms of tiredness, lethargy, dislike of cold weather (beyond the experience of others in the same environment), weight gain, depression, **constipation,** hoarse voice, menorrhagia (excessive menstrual flow), muscle pain and slow mental activity.

The signs may include bradycardia (slow pulse rate) dry skin and hair, slow relaxing reflexes, peripheral neuropathy, non pitting oedema around the eyes, hands and feet. There may be swelling of the thyroid gland (goitre) depending on the cause.

Biochemically, there is decrease in thyroid hormone - T4 and an increase thyroid stimulating hormone, TSH (from pituitary gland in the brain) in a normal functioning pituitary gland. The T3 portion of thyroid hormone may be unaffected.

In a normal functioning thyroid gland, thyroid hormones regulate metabolism (the sum of the physical and chemical changes occurring in tissues, which consists of anabolism -reactions that convert small molecules into large ones and catabolism where large molecules are converted to small ones). It increases basal metabolic rate (BMR), protein synthesis and neural maturity, sensitivity of catecholamines (adrenaline and noradrenaline), regulates protein, fat and carbohydrate metabolisms, proper development and differentiation of cells as well as stimulation of cells of human body.

In hypothyroidism virtually all body functions including CNS, heart rate, intestinal motility etc slow down. A decrease in intestinal motility (peristalsis) results in the slowing down of intestinal contents and consequent increased water absorption and therefore **constipation.**

Inability to control diabetes by diet or medication has already been dealt with above under diabetes as a cause of constipation. The facts are basically the same.

(c) Systemic Diseases

There are a number of systemic diseases (affecting the whole body systems) that are associated with **constipation**. The relationship is not direct but due to the general debility and weakness of the patient and also their multi organ involvement-the heart, kidney, brain, muscles and liver. Most of these organs (brain, muscles, heart and kidney influence peristalsis and therefore the delay in the prompt passage of colonic contents directly (brain, muscles) and indirectly (heart, kidneys). See above. Involvement of these organs by systemic diseases would affect their proper functioning and consequent **constipation.**

Some of the systemic diseases associated with **constipation** are listed below-

Systemic Lupus Erythematosus (SLE)
Scleroderma (systemic sclerosis)
Amyloidosis

Systemic Lupus Erythematosus (SLE) is a multi-system auto immune disease in which auto antibodies are produced against a number of auto antigens with the formation of immune complexes which deposit in organs such as the kidneys. It is more common in Afro- Caribbeans and Asians women (90 %).

It is a remitting and relapsing illness causing fatigue (tiredness), fever during relapses, weight loss, muscle pain, enlarged lymph nodes, hair loss, non infective endocarditis (interior heart inflammation)

We have already alluded to the importance of normal functioning of organs such as the kidney, muscles and general good health in the prevention of **constipation.** In SLE, as noted above there is impairment in the functioning of organs, particularly the kidney resulting in constipation. The general debility and sometimes confinement to bed and lack of physical activity are additional factors in the genesis of **constipation.**

Scleroderma (systemic sclerosis) is a systemic or generalised disorder of connective tissue which affects the skin and organs causing fibrosis especially of the lung, heart, and kidney. Females are affected 4 times more than men.

CREST syndrome is part of systemic sclerosis in which there is **C**alcinosis, **R**aynaud's, o**e**sophageal and gut dysmotility (disordered motility) **S**clerodactyly (stiffness and tightness of the skin of the fingers with atrophy of soft tissue and osteoporosis of the distal phalanges) and **T**elangiectasia (dilatation of terminal or small vessel of the skin in a particular part)

You would note that in CREST syndrome which is part of systemic sclerosis there is oesophageal and gut dysmotility. This dysmotility involves also the slowing down of colonic peristalsis which together with widespread systemic involvement of organs particularly the heart and kidney would combine to potentiate the resulting **constipation.**

Amyloidosis are a group of disorders characterised by extra cellular deposition of insoluble proteins in abnormal fibrillar form that resists degradation. They are usually divided into primary, secondary (due to diseases or illness) and familial amyloidosis.

Primary amyloidosis (light chain amyloidosis) is seen inpatients with myeloma, lymphoma, kidney disease, cardiomyopathy (heart disease), peripheral and autonomic neuropathy and gut macroglossia (enlargement of the tongue), perforation, obstruction and hepatomegaly (enlarged liver).

Secondary amyloidosis (Acquired Systemic Amyloidosis, Reactive Amyloidosis) occurs in chronic inflammation or infections, rheumatoid arthritis, inflammatory bowel disease, TB, bronchiectasis, osteomyelitis, and also affects the kidney, liver and spleen.

Amyloid is deposited in the brain of patients with Alzheimer's disease, type 2 diabetes and in haemodialysis.

It clear from above that amyloidosis of the kidney, heart, autonomic nervous system and the brain would cause **constipation** in the manners already described above. This is due to the involvement of the organs and effect on the gut generally and the colon specifically in the manner already described. **Constipation** may be exacerbated by general ill health and immobility which may be associated with this condition.

Chapter Six

Life Style, Diet and Miscellaneous Conditions and Constipation

We have so far dealt with more specific requirements to prevent constipation especially 3FEV factors in chapter one and have also looked at the mechanism of the causation of constipation; the slowing down of contents of the colon and consequent excessive water absorption resulting in SHIID.

There are certain conditions where despite adherence to 3FEV may still unfortunately end up with SHIID. In fact, these same problems apply to other conditions already dealt with in previous chapters. These chapters should enable the reader to consider other factors which may be responsible for SHIID despite keeping a clean sheet in respect of 3FEV. That is why I have tried to include the signs and symptoms of some of these conditions to enable the reader to realise when to consult their family doctor when they consider the possibility of an underlying condition or when there is a change in bowel habit.

I can not emphasise enough that some of these conditions are fortunately quite rare and the vast majority of cases of constipation or SHIID perhaps 80-90 percent are due to "normal" factors by simply not paying enough attention to 3FEV.

Unlike some of the previous chapters some of the conditions here are, in fact, quite straight forward and easy to spot and fix apart from the systemic diseases which require some in depth knowledge and paying particular attention to the symptoms.

I would like to look at this chapter under the following headings: -

1. Change in Routine
2. Suppressing the urge to open the bowel
3. Pregnancy
4. Stress
5. Drugs
6. Ageing
7. Food and Drink
   -Caffeine
   -Milk
   - Alcohol
8. Laxative overuse

1. Change of Routine

When we are used to doing things in a particular way and time, a change could make a great deal of difference. For instance if we normally go to bed at 10pm and get up at 6am if we start working nights, come back home at 9 am, try to sleep, get up at 8pm, prepare to go for work, then our so called "body clock" could be completely messed up. What of if we used to go to the toilet at 6am and at that time we are now at work, then when we do come home we have missed our usual toilet time as we had been working through out the night and now just want to sleep? This change of routine is likely to affect our "body clock" to the extent that **constipation** ensues due to this alteration in routine. Quite often the change doesn't have to be as noticeable as above, sometimes quite trivial changes like buying a new house with noisy neighbours and being woken up earlier than usual could trigger other changes and **constipation**. Some people appear to be more susceptible than others who can easily adapt to minor changes without their digestive system being affected.

There appears to be little doubt that the so called "body clock" or Circadian rhythm have some biological effects which are capable of altering our normal physiological processes when disturbed. The alteration appears to have more physiological disruption in some people than others.. This is exemplified by the "jet lag" phenomenon which occurs after a long flight which involves the crossing of different time zones. Many people experience changes such as disorientation, fatigue and sleeplessness because their normal routine has been badly disturbed not only by a long flight but because of prolongation of a normal day or night time and the body's normal physiological processes find it difficult to adjust and cope. This disruption could be serious to the extent that it affects alertness, appetite, hormone secretion and even temperature and undoubtedly also affects the digestive system-the bowels. The interplay of sleep disturbance, tiredness, disturbances in hormone secretion, stress and poor appetite are further factors in causing **constipation**. Fortunately not everyone would experience these changes to the same degree which may be explained by inherent factors and adaptation.

2. Suppressing the Urge to go to the toilet.

This is a common problem not only in children but even in adults. In children this is likely to occur because they are more interested in playing with their friends or busy with other things especially gadgets and toys. They would often ignore the call of nature and continue with what gives them pleasure at the time especially if their friends are around. By the time they finish what they were doing, that call of nature is no more and so the tendency is to continue with their activities and forget about it all. As such behaviour becomes repetitive, **constipation** becomes established with accumulation of faecal matter in the recto sigmoid region. The more the accumulation, the more the absorption of water and the harder the stool.

In adults this is more likely to occur because we would rather get home and use the toilet in which we are used to and is hygienic to our standard rather than the toilet at work, train, plane or city centre where we hesitate as we have no knowledge of who sat there last or before and how many people have actually used it that day before us. As we cannot attest to their standards of hygiene, would rather wait until we reach home. By so doing, the call of nature has completely gone. Sometimes apart from considerations of cleanliness we ignore the call of nature at work in order not to disrupt our routine, trend of thought, or fail to meet targets in today's competitive world. If the boss is around we would rather not give any bad impression as our promotion and prospects lie in being seen to be working hard and avoiding distractions.

There are times, when it is just laziness getting out of bed especially in a cold or unpleasant surroundings and at other times, like children, we would rather finish watching the final and rather exciting end of a television programme, soap, movie or sports.

By ignoring the call of nature a few times, the pattern becomes established, the sensitivity and recognition of the urge to defecate is diminished or virtually lost. The result is **constipation** as the neural and distension mechanisms in the rectum that would normally alert to the call of nature are diminished and fail. The situation is worsened particularly if such behaviour becomes repetitive.

3. Pregnancy is another physiological (normal) condition in which there is an increase in the "incidence" of **constipation**. This is not entirely unexpected as there are several changes in pregnancy which upset "normal" body mechanisms.

One of such changes is hormonal. During pregnancy there is an increase in progesterone which has effect not only on the uterus and foetal development but also on the gastrointestinal tract where it slows down gut motility (peristalsis). In addition, the growing foetus has a moderately "obstructive effect" on parts of the gut distorting the regular luminal flow of contents. These combined hormonal and "obstructive effect" account for about a 25 percent increase in **constipation** during pregnancy.

4. Stress

It has long been known that people become constipated when they are stressed or have some mental health problems. Stress and anxiety interfere with our normal routine. Often when people are stressed they tend not to eat, sleep, interact, exercise or get out of bed early. These limitations in their activities and healthy eating habits as pointed out in chapter one is a potent cause of **constipation.** Review the 3FEV- "three female examples of virtue" (in chapter one) which are less likely to be adhered to during periods of stress.

We must also recognise that stress is a neurological or central nervous system problem. Gut motility is a function that is governed by the CNS as well as enteric or autonomic nervous system. We have seen in earlier chapters how interference in these systems can cause the slowing down of gut motility and thus cause **constipation.**

It is also pertinent to point out that people who are stressed tend to indulge more in alcohol which has a diuretic/dehydrating effect leading to **constipation** A number of drugs prescribed for stress and related problems cause **constipation** and stressed people are also more likely to indulge in caffeine which as will be seen later is another cause of constipation.

Lastly it is pertinent to point out that our body systems including our digestive system respond to mental health in different ways. Anxiety and stress can cause tachycardia (fast pulse and heart rate) sweating (skin), tachypnea (fast breathing). These are autonomic nervous system responses due mainly to increased catecholamine and sympathetic activity. As pointed out earlier the gastrointestinal tract tend to respond in a negative manner to sympathetic stimulation or activity by decreased motility (peristalsis) and inhibition of ano-rectal sphincters. The results of these responses lead to **constipation.**

5. Drugs or Medications

These will be dealt with in details in a separate chapter but suffice it to say here that they are a number of drugs which cause constipation.

These include-

(a) narcotics (opioid) which are drugs like morphine, codeine or codeine containing drugs like Co-Coda mol.

(b) Anticonvulsants like phenytoin, and carbamazepine.

(c) Antidepressants like amitriptyline

(d) Haematinics such as iron supplements

(e) Antihypertensive and heart medication-Calcium blocking agents like nifedipine, amlodipine, diltiazem, verapamil, etc

(f) Aluminium containing antacids like Aluminium hydroxide, Hydrotalcite, Asilone and Maalox plus.

(g) Diuretics (medications that increase urine production and excretion) Used to relieve oedema and treat hypertension, for example Bendroflumethiazide.

6. Ageing.

Age is an important factor in **constipation** and will be dealt with separately in another chapter because of its importance and how it can be avoided and managed successfully. Suffice it to say here that the body processes slow down at this stage in life and so is the bowel. To add insult to injury, many elderly people do very little exercise: some are virtually bed-ridden due to age related diseases like Parkinson's disease, multiple sclerosis (MS), endocrine and metabolic diseases, strokes, heart failure, chronic obstructive pulmonary disease (COPD) etc, all of which are associated with constipation directly or indirectly. Besides, they are often on multiplicity and varied medications, which as seen above, some may be responsible for **constipation.**

This should not be the case as age is no barrier to exercise and medications should be regularly reviewed and discontinued if no longer necessary. Age related conditions should be treated or well managed and above all the principles of 3FEV (3 Female Examples of Virtue) introduced and adhered to. In this way a general well being is achievable and constipation avoided or reduced to the barest minimum.

7. Food and Drink
   - caffeine
   - milk
   - alcohol

The food we eat and drinks we have unfortunately do cause constipation depending on the type and our system's susceptibility or intolerance to such food and the general factors that normally moderate or cause constipation. We should not neglect our 3 FEV as these provide the best way of prevention of constipation.

Caffeine is contained in a number of drinks especially coffee and tea. Coffee contains a good quantity of caffeine. Some people experience noticeable constipation attributable to drinking coffee; others appear not to notice any untoward effect. As stated above some people appear to be more prone or susceptible to certain drinks and food than others.

Caffeine is a stimulant which has some effect on the central nervous system. Such stimulation may have a sympathomimetic effect (mimicking the sympathetic autonomic nervous system). As noted in earlier chapters, the sympathetic nerve inhibits colonic motility and constricts ano-rectal sphincters. This effectively means inhibition of peristalsis and sphincteric action which is **a constipating** effect.

Stimulation of the CNS sympathetic axis is also **a constipating** effect in the same vein for similar reason.

Notably also, tea and coffee tend to keep many people awake all night not just for the stimulant effect, but for another reason- the diuretic (an agent that increases the amount of urine excreted) effect. Excessive urination dehydrates and in order to maintain fluid balance, the body "steals" water from where ever it can. One unfortunate area that suffers the effect of this "theft" is the colon with "dire consequences" -**constipation**. There is increased absorption of water from colonic contents resulting in dry hard faeces and the conditions for SHIID (**constipation**) are met.

Milk. Like coffee, many people have no constipating effect due to milk and milk products; but others do.

It has been observed that many babies develop constipation with cow milk. Dr Alan Greene of DrGreene.com has suggested that this effect could be due to the milk protein and that these babies may be lactose intolerant and that switching to soy milk could dramatically improve their constipation.

In a study done at the University of Italy in 1998, 65 children with chronic constipation, who had been prescribed laxatives and even changes in diet which had failed to stop their constipation, were randomly distributed into 2 groups and were "blindly" given either cow's milk or soy milk. The 2 groups were given the two types of milk for a period of 2 weeks, then stopped for one week. They were then swapped and the children who were on cow's milk were then given soy milk and vice versa.

An analysis showed that whilst the children were on soy milk, 68 percent experienced no constipation. The inference is clear that these children who were lactose intolerant suffered the untoward effects of cow milk with resulting constipation.

Dr. Alan Green has suggested the use of other forms of milk-almond milk, hemp milk, oat milk, all fortified with important vitamins, minerals and vitamin D.

The reason for advocating other types of milk is easy to see. In the University of Italy study, only 68 percent of children benefited completely from soy milk and their constipation stopped. What of the remaining 32 percent?

It would appear that the remainder of the children who did not benefit from soy milk could benefit from the other types of milk suggested above or more appropriately there is another unidentified factor other than milk that is causing their constipation. Such consideration borders on a confounding factor ("a relationship between the effects of two or more causal factors observed in a study or set of data that is not logically possible to separate the contribution of any single agent to the observed effect"). This is a public health or **epidemiological** consideration (relating to the "study of the distribution and determinants of health related states or events in specified populations and application of this study to control of health problem")

It is very likely that the constipation these children suffered was due to the cow's milk and their lactose intolerance to the milk but more studies may be required to find out why 32 percent failed to respond to soy milk unless a good majority respond to other types of milk.

The constipating mechanism is also not clear cut. It looks likely that cow's milk or more specifically lactose has constipating effect whilst, in fact, some other children have quite the opposite effect of diarrhoea, vomiting and abdominal pains. It is suggested that lactose may have a constricting effect on gut muscles of some people which slows down gut contents resulting in **constipation.**

Alcohol.-This has a constipating effect because alcohol acts as "diuretic" (see above)

You would have observed that there is frequency and increased volume of urine that is disproportionate to the quantity of alcohol consumed compared to drinking water. This is due to its diuretic effects which makes the kidneys excrete more urine. This dehydrates and the body immediately sets its compensation mechanism in action to restore normal volume and the colon is again an easy prey. This leads to more absorption of water from colonic contents and the resulting hard stool and consequent **constipation.**

Laxatives over use. This is one cause of constipation especially with over the counter laxatives. Unfortunately some people have the notion that they have to open their bowels a certain number of times a day. The truth is that there is a wide variation ranging from 1-3 times a day up to 1-3 times a week. Individuals differ markedly.

There is therefore a tendency to become dependent on laxatives with a propensity to increase the dose in order to have the same effect and when the laxative is stopped, constipation results because the system has become used to taking laxatives in order to have bowel motion.

CHAPTER SEVEN

Medications that Cause Constipation

Medication is one of the most common causes of constipation. The list is virtually inexhaustible and ranges from commonly used painkillers like Codeine, iron supplements for anaemia to medication used in hypertension, heart disease, mental health problems and Parkinson's disease. Some of the drugs act centrally in the central nervous system and slow down bowel motility through this action. Others exert diuretic effect (excessive excretion of water by the kidneys) thereby causing the colon to absorb back more water than usual to try and counterbalance the reduction in fluid volume. The result is the passage of hard stool, with straining, small in quantity, infrequently and with difficulty (SHIID- see chapter one). Other medications like ferrous sulphate which are haematinics used to supplement iron stores in the body in cases of iron deficiency anaemia and aluminium containing antacids exert their constipating effect locally.

There are other medications which cause constipation in which no simple explanation of the mechanism or specific effect leading to constipation can be found. Many of such cases are peculiar to some individuals whilst others report no constipation. In some of such cases there may be some other cause of the constipation but the patient attributes it to the medication.

It is important that whatever medication that causes constipation, the principles of 3FEV is followed as much as possible. Plenty of fruits, fibre, fluids, vegetables and exercise are very likely to stem the tide of constipation. Don't forget the "three female examples of virtue". It is imperative therefore that medications are not discontinued merely because of constipation if they are prescribed for a specific and important medical condition but to ease or stop the constipation by adopting the 3FEV solution. The alternative to this is, of course, to go to the family doctor to review the medications and prescribe possible alternatives or discontinue if the conditions for which they were prescribed no longer exist or have become so negligible that continuation of the medication would make no significant difference to the health of the individual but constipation itself has become too big a problem to handle than the medical condition. Such cases are rare as medications are given for specific medical conditions that require treatment but constipation is quite often a passing inconvenience and there are remedies embodied in 3FEV to deal with it to the extent that it should no longer constitute a problem.

Are there always alternative medications that have no constipating effect? The simple answer is no. The reason for this is that many medical conditions have specific routes of "cure" or in effecting amelioration of symptoms. For this reason, most of the effective medications tend to follow the same pattern and quite often the prescribed drugs may be the most effective. Even when alternatives exist, they may well be the weaker versions of the potent medication.

When treating moderate/ severe pain, the doctor would rather prescribe Co-Codamol which causes constipation due to the fact that it combines codeine and Acetaminophen (Paracetamol) than the weaker alternative Acetaminophen only with virtually no constipating effect. The same applies to severe pain. Prescription of strong potent analgesia like Morphine which causes constipation would be much preferred to weaker painkillers which would not cause constipation but unable to control severe pain. Sometimes or indeed quite often some of us would be ready to accept constipation as the price to pay for effective pain control.

There are other medications used for mood disorders like Imipramine and amitriptyline and medications used in psychosis and schizophrenia such as Olanzepine, Haloperidol and Risperidone all of which act centrally in the central nervous system (CNS) and cause constipation and there are no alternative medications that act peripherally that would obviate constipation. These medications therefore have to be prescribed despite their constipating effect. The alternative of not prescribing the medication merely because of constipation, can not even be considered as the benefits of treatment far outweigh the side effect of constipation.

Medications that cause constipation can broadly be divided into the following groups

1. Drugs that act mostly locally in the gastrointestinal tract.
2. Drugs that have diuretic effect
3. Drugs that act centrally on the central nervous system (CNS)
   (a) Pain relief medication-Opioid Analgesia.
   (b) Drugs used for Parkinson's Disease.
   (c) Drugs used for Schizophrenia and psychosis
   (d) Drugs used for stabilization of mood disorders -depression and anxiety.
   (e) Cough suppressant with Opioid
   (f) centrally acting appetite suppressant

4. Drugs used to treat diarrhoea.
5. Antihypertensive medications.
6. Drugs used to treat high cholesterol (hypercholesterolemia).
7. Medications used to treat peptic ulcer.
8. Drugs used to treat iron deficiency anaemia.
9. Drugs used to alter gut motility and for other uses (antimuscarinics).
10. Herbal Medication- St John's wort.
11. Anticonvulsant Medications (antiepileptic medication).
12. Anti inflammatory medication

All these medications have a variety of actions that tend to cause **constipation**; some more than others. Whilst some people may have constipation with some of these medications, there are also others who are unaffected. This may be due to individual susceptibility but more likely due to the adoption of non constipating life style by some but not those affected. As already discussed, fruits, fibre, fluid, exercise and vegetables (3FEV) are likely to ameliorate the effects of any constipating drugs and this may explain why some people do not suffer constipation as side effect whilst on these medications or at least do so to a lesser degree than others.

I like to look briefly at some of these medications and when possible elucidate the likely mechanisms of their constipating effect.

(1)   Drugs that act mostly locally on the gastrointestinal tract

An example of this is Aluminium containing drugs used in the treatment of peptic ulcer. This is typified by Aluminium Hydroxide.

Aluminium Hydroxide is relatively insoluble in water and this property may be responsible for its constipating effect.

(2)   Drugs that have diuretic effect.

Diuretic medication promotes the excretion of large volumes of urine. They are used in the treatment of oedema, hypertension and pulmonary oedema due to left ventricular failure.

Their **constipating** effect is due essentially to the excessive loss of body fluid and as already explained, the body responds by absorption of more fluids to replenish the loss and the colonic contents is one area where more fluid is reabsorbed. The result is the production of hard faeces and there may be straining, infrequency, insufficiency and difficulty (SHIID)

Diuretics are divided into different groups based on their mode of action

(a)   Thiazide diuretics eg bendroflumethiazide (bendrofluazide) are moderately potent diuretic and used mainly in the management of mild hypertension and mild to moderate heart failure either alone or in combination with other forms of medication.

They exert their diuretic effect by inhibiting sodium reabsorption at the beginning of the distal convoluted tubules of the kidney. By this action more sodium is lost in the urine accompanied by water thus producing large urine volume and frequency. Sodium (like common salt, table salt) is associated with hypertension and lowering its plasma level helps to reduce the blood pressure and plasma volume resulting in the shedding or unloading of excess fluid thus relieving oedema

(b)   Loop diuretics like Furosemide (Frusemide) inhibit reabsorption from the ascending limb of the loop of Henle in the renal tubules. They are more powerful diuretics and used for pulmonary oedema due to left ventricular failure and relieves breathlessness. They may also be added to other antihypertensive treatments.

(c)   Potassium sparing diuretics like Amiloride and Aldosterone antagonist like Spironolactone, unlike the above help to retain Potassium which are lost during treatment with Thiazide or loop diuretics along with sodium. They are therefore often given with them if there is any risk of low potassium (hypokalaemia)

(d)   There are also Potassium sparing diuretics combined with other diuretics-
Co-amylozide
Co-amilofruse
Co-triamterzide
Co- flumactone
Lasilactone

(e)   Mannitol is a separate group of diuretic in the treatment of cerebral oedema.

The mechanism of causation of **constipation** in all the diuretics are essentially the same as explained above. It is important to emphasise that not all the people who take these diuretics will have constipation: other factors embodied in 3FEV come into play and constipation can be avoided.

3.   Drugs that act centrally on the central nervous system cause **constipation** through their interference in the nerve connections that promote increased bowel motility and thus peristalsis. That is the basic mechanism which has already been explained in previous chapters in this book.

Briefly these drugs can be classified as follows: -

(a) Pain relief Medications. These are Opioid Analgesics like the following: -

Codeine
Co-coda mol
Dihydrocodeine
Pethidine
Morphine
Fentanyl
Methadone
MST
Oramorph
Tramadol

There are some of the commonly used and prescribed opioid analgesics

(b) Drugs used for the treatment of Parkinson's Disease: -

Bromocriptine
Rotigotine
Levodopa
Tolcapone
Entacapone
Procyclidine
Trihexyphenidyl
Selegiline Hydrochloride
Rasagiline
Ropinirole
Pergolide

(c) Drugs used for the treatment of schizophrenia and psychosis: -

Haloperidol
Chlorpromazine
Flupentixol

Risperidone
Pericyazine
Prochloperazine
Clozapine
Amisulpride
Quetiapine
Olanzapine
Zuclopenthixol Decanoate (Clopixol)

(d)

Drugs used for stabilization of mood disorders-Anxiety, Depression and Attention Deficit Hyperactivity Disorder (ADHD)

Citalopram
Escitalopram
Clomipramine
Imipramine
Amitriptyline
Doxepin
Nortriptyline
Dosulepin Hydrochloride
Phenelzine
Citalopram
Venlafaxine
Reboxetine
Atomoxetine
Modafinil

(e) Cough suppressants with Opioids

Codeine Phosphate
Pholcodine

(f) Centrally acting appetite suppressant.

This acts centrally and causes constipation in the same way as other drugs that affect the central nervous system and with its connection to the peripheral nervous system and the colon causes constipation as already explained.

An example of this is Sibutramine Hydrochloride.

4.   Drugs used to Treat Diarrhoea

There are 2 main reasons why anti diarrhoea drugs cause constipation. The main reason is that diarrhoea and **constipation** are at opposite ends of the spectrum with diametrically opposed pathologies. Treatment of diarrhoea is the first step in the causation of constipation and this may be dose or time dependent. It is important to strike a balance so that successful treatment of diarrhoea does not progress to the causation of **constipation** which is an undesirable effect of the treatment. It is therefore important to keep to prescribed or recommended dose and to stop treatment when the aim is achieved.

The second reason why anti diarrhoea medication causes **constipation** is the fact that constipating medication like the Opiods are used to treat diarrhoea. The conversion of diarrhoea to **constipation** is then a question of observing dose and timing as explained above.

It is important to note also that despite this propensity, the problem can easily be overcome by the proper use of 3FEV. See chapter one. The importance of the 3rd F (Fluids) can hardly be overemphasised in the treatment of diarrhoea as there is much fluid and electrolytes lost in the stool and should be replaced. This is particularly important in the elderly and infants. This fluid loss is the first step in the causation of **constipation** due to the formation of hard stool

There are 2 main categories of drugs used in the treatment of diarrhoea which overuse or prolonged use can cause constipation.

(a) Adsorbent and bulk forming drugs.

Adsorbent drug like kaolin -used for chronic diarrhoea.
Bulk forming drugs like methylcellulose, sterculia and ispaghula which are used for diarrhoea caused by irritable bowel syndrome.

(b) Antimotility Drugs are used mainly in acute diarrhoea in adults and include drugs like-
Loperamide
Co-phenotrope,
Codeine Phosphate

A mixture of (a) and (b) Kaolin and Morphine Mixture is also available

5   Antihypertensive and Angina Drugs-

Calcium -channel drugs which interfere with inward displacement of calcium ions.
Though their actions are mostly in the cardiac and vascular cells where they depress the contraction of heart muscle and vascular smooth muscle but some are not so discriminate and specific in action: they may depress smooth muscles of the gastrointestinal tract as well. This effect may also be mediated through the depression of vascular smooth muscle supply to the intrinsic and extrinsic innervation of the colon. The depression of colonic smooth muscle reduces colonic motility thereby reducing peristalsis leading to **constipation.**
These drugs include-

Verapamil
Amlodipine
Felodipine
Lacidipine
Nifedipine

Centrally acting antihypertensive, Clonidine, also causes constipation through the same mechanism as medications acting on the brain as already dealt with above.

(6)   Drugs used to treat high cholesterol (hypercholesterolemia)

These are medications that are bile acid sequestrants such as colestyramine, colestipol and colesevelam. They cause **constipation** because they are not absorbed but act by binding bile acids preventing their reabsorption which promotes conversion of cholesterol to bile acids in the liver. At the extreme end, intestinal obstruction has rarely occurred because of its mechanism of action.

It should be noted that the commonly used cholesterol reducing medication, the statins have a different mechanism of action and so constipation is not usually a problem.

7.   Medications used to treat peptic ulcer.

Sucralfate

Sucralfate is a complex of aluminium hydroxide and sulphated sucrose which acts locally by protecting the mucosa of the stomach and duodenum from acid attack. Aluminium Hydroxide as already explained above is relatively insoluble in water and this property may be responsible for the **constipation.**

8.   Drugs used to treat iron deficiency anaemia-

Ferrous Sulphte
Ferrous Fumarate
Ferrous Gluconate

The mechanism of the causation of constipation by iron tablets is not entirely clear. It is thought that being metallic, this may slow down neurological activity and thus reduce gut motility and peristalsis causing constipation. Another school of thought is linked to its effect of marked changes in intestinal flora which could predispose to constipation..

Whatever the actual mechanism, it is worth noting that not all the people who take iron tablets experience constipation. This is likely to be dose related in addition to individual susceptibility: a reduction in dose may reduce any constipating effect. There is in addition the necessity to increase fruit, fibre, fluid and vegetable intake as well as increased physical activity -exercise (3FEV) as a means of combating any constipating effect of iron tablets.

9.   Drugs used to alter gut motility and other uses below (antispasmodics)

These drugs reduce intestinal motility - some by direct relaxation of intestinal smooth muscle and which account for their constipating effect. They are often used in the management of irritable bowel syndrome (IBS) and diverticular disease. There are other uses of antispasmodics out side the gastrointestinal tract. They are also used in arrhythmias, asthma, airway disease, motion sickness, Parkinson's disease, urinary incontinence and as antidote in organo- phosphorus poisoning. There are also other medications that have muscle relaxing effect and cause constipation.

They include -

1. Antimuscarinics
   Atropine Sulphate
   Hyoscine Butylbromide
   Propantheline Bromide

2. Antispasmodics (Directly relax intestinal smooth muscles and may cause constipation)
   Alverine
   Mebeverine
   Peppermint Oil
   Antimuscarinics used in asthma and chronic obstructive pulmonary disease (COPD) include-
   Ipratropium Bromide (Atrovent)
   Tiotropium

Antimuscarinics used in urinary incontinence. There is usually involuntary detrusor contractions which cause urgency, urge incontinence with frequency and nocturia. Antimuscarinic drugs are used to reduce the contractions and increase bladder capacity. These symptoms are often seen in overactive bladder syndrome. Their side effect may cause **constipation**. These drugs include-

Oxybutynin-which has a direct relaxant effect on urinary smooth muscle.
Tolterodine
Darifenacin
Fesoterodine
Solifenacin
Propiverin
Trospium

Although antimuscarinics used in urinary incontinence act mostly on bladder muscle, they also have some relaxing effect on gut muscle. Relaxation of colonic muscle reduces motility and leads to **constipation.**

10   Herbal Medications- St John's Wort

Commonly used in many countries for depression, anxiety, tiredness, menopausal symptoms, loss of appetite, insomnia, stress, palpitations, attention deficit hyperactivity disorder (ADHD), obsessive compulsive disorder (OCD) and seasonal affective Disorders.

They are not usually prescribed by doctors but patients buy them usually in herbal shops and are said to be efficacious in some of the conditions above.

St. John's Wort is said to blossom about June 24th the birthday of John the Baptist and so named after him.

Some researchers have done some work on it and isolated the chemicals hyperforin and hyperacid which act as chemical messengers in the central nervous system (CNS) and regulate mood changes.

Like other drugs that act on the central nervous system they have the tendency to affect gut motility adversely by slowing it down (see earlier chapters) This reduces peristalsis and speed of colonic contents resulting in **constipation**.

11. Anticonvulsant Medication (antiepileptic)

Although anticonvulsant medications act centrally and cause **constipation** in the manner already discussed, I have decided to highlight them again as a separate group. Because of their effect on the CNS, some are responsible for both constipation and diarrhoea and others either of the two symptoms. Some of these drugs are also used in the treatment of mental health problems shown above.

These medications include: -

Pregabilin
Gabapentin
Carbamazepine
Rufinamide
Phenytoin
Lacosamide
Oxcarbazepine

12.- Anti inflammatory and anti rheumatic medications

These may cause diarrhoea or **constipation**. These symptoms are common with suppositories due to rectal irritation and tablets due to local effect which causes discomfort due to damage to intestinal mucosa (ulceration) and may lead to bleeding or even perforation in prolonged use which should be discouraged. There is no certainty whether the damaged intestinal mucosa is responsible for the constipation which some patients complain of. Some patients actually experience diarrhoea so the gastrointestinal side effects are quite variable. Indomethacin may cause intestinal stricture which limits the free movement of intestinal contents resulting in **constipation.**

They include:

Ketoprofen (Orudis)
Diclofenac Sodium
Indomethacin- which may cause intestinal stricture and suppositories rectal irritation.

It is clear from the above examples that medications are responsible for a sizable number of complaints of constipation. It is worth looking up the list of these medications if constipation is experienced whilst on medication despite adherence to the 3FEV which should reduce or abolish constipation.

CHAPTER EIGHT

Ageing Process and Constipation

Older people tend to suffer more from constipation. There are several reasons for this which will become clearer in this chapter.

We are now living longer than our grandparents and this is predicted to keep on increasing year in and year out and has been due mainly to the global elimination of many diseases, better medical care, improvements in social services, education, cleaner environment, more focus on research and utilization of the research for the manufactor of medications, equipments and better treatment options and of course better nutrition and exercise.

Despite all the advances, if we survive the vicissitudes of life and advance to adulthood, then the ageing process inevitably takes over. This process causes changes to our mental and physical health which fortunately can virtually be resisted for several years if we adopt the right options in lifestyle, diet and exercise. We often hear people say things like; "he is 80 years of age but has the mental and physical attributes of a 60 year old" You don't need cosmetic surgery to achieve that but modification of lifestyle and better nutrition.

However despite the best care and nutrition, the inevitable eventually happens in the "Seven Ages of Man" **and in the words of William Shakespeare**:

"All the world's a stage,
And all the men and women merely players:
They have their exits and their entrances;
And one man in his time plays many parts,
His acts being seven ages.
At first the infant,
Mewling and puking in the nurse's arms.
And then the whining school-boy, with his satchel
And shining morning face, creeping like snail
Unwillingly to school"
- - - - - - - - - - - - - -
"Last scene of all that ends this strange eventful history
Is second childishness and mere oblivion,
Sans teeth, sans eyes, sans taste, sans everything".

That is the inevitability and reality of ageing expounded succinctly and poetically by William Shakespeare but this can now be resisted for several years for reasons stated above.

Although constipation is a greater problem in old age due to the factors to be outlined later, it is quite possible to ameliorate them by adopting healthier lifestyle which will usually require some effort and determination.

The ageing process is the gradual deterioration of youthfulness: there is decline or reduction in muscle mass and strength, hearing, eyesight, health as we knew it and cognitive abilities which we had taken for granted over the years. The net effect is that our organs and systems not only experience a decline but take longer to achieve the desired outcome. Our digestive system is not exempt from this global degeneration and **constipation** is one manisfestation. Fortunately we can delay this process for several years by adopting the right measures.

Apart from prompt attention to health issues, prevention is usually better than cure. We are the products of our lifestyles and the food we eat.

I would again like to emphasise the 3FEV (Fruits, Fibre, Fluids, Exercise, Vegetables) in chapter one. This is the bedrock in overcoming constipation and becomes even more important in old age as a result of the deterioration and decline in our body systems and the weakness of muscles including the gastrointestinal tract which affects the force and speed of motility of intestinal contents of which stasis results in **constipation.**

There are other very important factors which affect our health generally at all ages and become more important in old age. These factors can be prevented by and are stated and remembered as follows: - cessation of smoking **(Cigarettes),** reduction in **Salt, Sugar** and **Alcohol intake** which together make up **CISSA**. (Pronounced Caesar) They are better represented as follows: -

CIgarettes
Salt
Sugar
Alcohol

CISSA is an important health message, Though not a direct cause of constipation in some cases but impacts on general ill health which is responsible for many cases of **constipation** which will become obvious below.

CISSA elements individually have direct relationship with cancer, cardiovascular disease, hypertension, strokes, obesity, diabetes and mental health which we have seen in previous chapters and will see again later in a different light about their importance in the genesis and propagation of **constipation**. These conditions may be age related, hence their significance in this chapter. However, it must be emphasised that they affect virtually all age groups from adolescent to old age and their inclusion here is mainly for convenience only so the health message is for all.

The Effects of CISSA on Health and Constipation

1. ## CIGARETTES

(CI). Smoking is one of the most important factors in the causation of ill heath. It is a major cause of **cancer** especially cancer of the bronchus, lung, pancrease, bladder, stomach, colon, rectum, liver, kidney, ureter, and oesophagus etc.

In the **respiratory system**, smoking causes chronic obstructive airway disease (COPD) which includes emphysema and chronic bronchitis, also smoking worsens asthma. We have already noted above, the cancers caused by smoking in the respiratory system.

In the cardiovascular system (CVS) cigarette smoking causes coronary heart disease (CHD), heart attacks, strokes and peripheral vascular disease which can result in severe pain in the legs and amputation or surgery with limitation of movements including physical activities and exercise.

It is evident that conditions such as cancer, COPD, heart attack, peripheral vascular disease with or without amputations are responsible for severe invalidity, immobility and prolonged confinement to bed. Immobility and invalidity restrict physical activities and exercise which are responsible for several cases of **constipation.** In COPD, exercise tolerance is severely diminished.

Cigarette smoking is also implicated in cancers of the stomach, pancrease, colon and rectum. In the latter two cases (cancer of the colon and rectum), obstructive lesions may present as **constipation.**

The important health message about cigarette smoking is total cessation. This is not usually easy as nicotine is highly addictive. Stopping smoking thus requires great effort and determination to kick this unprofitable and unhealthy pastime. Professional assistance may be required to do so.

2 **SALT (S)** or sodium chloride is present in most food items including naturally occurring foodstuff like fruits, vegetables, potatoes, meat, fish etc. In order to improve taste, it has become customary for food manufacturers to add salt to processed food like crisp, bread, cereals, meat, also restaurant, canteen and take-away food.

The recommended salt intake is 5-6 gm per day but because of added salt in food that we prepare at home and hidden salt in processed food, many people unknowingly exceed this limit 2-3 fold.

Whilst a little salt is essential to the body, excess salt causes **hypertension**, **strokes** and **heart disease** and has also been implicated in the causation or exacerbation of kidney disease, osteoporosis and stomach cancer.

Cutting down on salt intake is not very easy because we have become used to the taste in the food we eat but it is much easier than cessation of cigarette smoking which is addictive. Salt is not addictive and only requires gradual reduction of intake and avoiding processed food with too much added salt which should usually be shown on labels and packaging. Unfortunately that is not always the case.

Reduction in the amount of salt in processed food is **a very important health message which can only be tackled by government legislation requiring food manufacturers to display on the label the amount of salt and placing strict limits on the amount of salt, if any, which can be added to processed food**. Alternatively the government should enact legislation to include **health warning** as in cigarettes about the hazardous effects of salt on health. A combination of the two legislations is my preferred option.

The emphasis on salt reduction and government legislations have been emphasised here because too much salt in our diet and processed food "is a silent killer". This message is for all governments globally and there is no need in naming countries again: having already done so at the beginning of this book. A full discussion with stakeholders could precede legislation and in some cases obviate the need for such legislation.

A full discussion of the effect of salt on health is not a digression. We have already alluded to its effect in the causation of **hypertension, strokes, heart disease** and immobilisation in the causation of **constipation.** I have only added other health dimensions and their concordance with **constipation**.

3. SUGAR (S)

Excess sugar is an important cause of ill health. There are some people who report having "sweet tooth". Like salt, sugar is not addictive in the true sense of the word. Taking too much sugar in your tea, coffee or cereal is just a habit which with determination can be broken. In many cases adding sugar to food and drink can be avoided all together because sugar is naturally present in carbohydrate food.

High blood sugar level (hyperglycaemia) is responsible for **diabetes.** It is worse in people who are already diabetic or have a propensity to diabetes through family history of diabetes.

High blood sugar level or **diabetes** is responsible for **damage to blood vessels, strokes, coronary heart disease**, **heart attacks**, **kidney disease, renal failure**, visual impairment or blindness. It is common cause of **peripheral vascular disease, ischaemic limb and gangrene requiring amputation.**

We have already seen the effects of these conditions in the causation of **constipation**. Briefly, let me mention some of them again. Diabetes causes neuropathy which affects colonic autonomic nervous system and if the balance shifts in the direction of the parasympathetic nerve, constipation ensures and, of course, the extrinsic afferent and efferent nerves of the colon are also affected. Strokes have CNS sequelae which result in **constipation.** We have already noted the morbidity and immobility associated with these conditions which make physical activity in general and exercise in particular impracticable. That is a prerequisite for **constipation.**

Like salt, should the government also not legislate against the adding of too much sugar to processed food? Surely, sugar content of all processed food should be recorded on the label and packaging to warn consumers of the amount of sugar in the drink or food. Large Multinationals who produce carbonated soft drinks should show the sugar contents on the cans and bottles of these drinks and carry health warning of the effects of sugar. That is where government legislation comes in. The government may not have the right (still it may have that right) to legislate against the sale or marketing of food and drinks that contain too much sugar or salt but has the right to legislate for manufacturers to show the amounts of sugar or salt in these products and like cigarettes carry some health warning so that individuals can make informed choices.

4. **ALCOHOL** (A)

Alcohol has been used in society since biblical times. The problem with alcohol is the addiction and dependence with regular use and the tendency to increasing use in frequency and quantity. Alcohol causes several health problems including **liver** cirrhosis**, hypertension, cancer** especially of the mouth, neck and throat, **heart attack**, irregular heart beat, fatigue, sleep and sexual problems as well as **depression** and **obesity.**

We have already talked about the relationship of these health problems and **constipation** in previous chapters.

Dependence and addiction present special problems both to the individual, family and society. Drink- driving (driving under the influence of alcohol) is a special problem to society.

Accidents wreck lives and cause serious injuries which require hospitalisation and immobility and the resulting lack of physical activity is an important factor in the causation of **constipation**.

The NHS recommends that men should not regularly drink more than 3-4 units of alcohol a day (21-28 units a week).

Women should not regularly drink more than 2-3 units of alcohol a day which is 14-21 units a week.

Drinking alcohol during pregnancy is not regularly encouraged but 1-2 units once or twice a week may be okay.

We have seen that the 4 elements of **CISSA**- Cigarettes, Salt, Sugar and Alcohol are responsible for virtually identical and very important health problems of **D**iabetes, **I**schaemic heart disease, **S**trokes, **C**ancer, **H**ypertension, and **Obesity (DISCHO)** all of which impact adversely on the bowel and cause **constipation**

These conditions can best be remembered as follows-

Diabetes
Ischaemic heart disease
Strokes
Cancer
Hypertension
Obesity

They are best remembered as **DISCHO** (pronounced disco)
You would also note that disobeying the rules of CISSA causes DISCHO.
Conversely and more importantly obeying the "laws" of CISSA stops DISCHO in its track.
We have seen the relationship between CISSA, DISCHO and **Constipation.**
Having seen the establishment of this relationship we can look more broadly at the age related conditions and **constipation** including those already mentioned above but in a general context.

There are several age related diseases and conditions which make senior citizens particularly susceptible to constipation.

Heart, kidney, muscle diseases, hypertension, strokes, hearing, poor vision, prostate problems, bowel problems, cancer-particularly bowel cancer are some of the problems associated with ageing.

There are many others including osteoporosis, overweight, depression, cognitive problems, diabetes, chest problems, poor appetite, dementia, gout, Alzheimer's disease, immobility, high cholesterol, high risk of falls, sedentary lifestyle, loneliness, social isolation and polypharmacy.

I shall subdivide these problems as follows and see how they relate to **constipation: -**

1. Age related diseases that cause constipation
2. I mmobility and sedentary lifestyle
3. Polypharmacy
4. Cancer particularly bowel cancer
5. Higher Risk of falls

1 Age related diseases and conditions that could predispose to constipation.

There are many diseases which I have already discussed above or in earlier chapters which predispose or actually cause constipation. Most of these diseases are age related. Prompt recognition, diagnosis and treatment should be the norm but quite often they are unrecognised or ignored until advanced stages when very little can be done. The predisposition to constipation may be direct or indirect and in the latter case may be due to the medication used in the treatment of the particular condition. Some of these conditions include the following: -

Heart disease
Kidney disease
Arthritis
Hypertension
Prostate cancer and benign prostatic hyperplasia
Oseoporosis
Depression
Anxiety
Cognitive problems
Dementia and Alzheimer's disease
Breathing problems especially COPD
Lack of social integration
Lung cancer
Gout
Strokes & TIA (Transient Ischaemic Attack)
Hypercholesterolemia
Diverticular Disease
Parkinson's Disease
Obesity and weight problems
Poor appetite
Diabetes
Hypothyroidism
Scleroderma (Systemic sclerosis)

We have seen in chapters 5-7, that some of the above conditions cause **constipation** due to their effects on the gastrointestinal tract, others to their effect on the heart, kidney and brain which directly or indirectly affect colonic motility adversely and thus the propulsion of colonic contents which is slowed down. Some of the age related conditions like poor appetite make the eating of a balanced diet difficult and virtually impossible to effect 3FEV resulting in **constipation.**

Mobility is important in prevention of **constipation** and this is particularly important in the elderly who may not be engaging in regular exercise. Breathing problems, lack of social integration, dementia, lung cancer, strokes, Parkinson's disease, scleroderma, arthritis and osteoporosis are some of the age related conditions that restrict mobility either directly or due to general debility. These will be dealt with separately under different headings.

We have seen the effect of certain medications in the causation of **constipation** in chapter seven. We also have to recognise that the treatment of most of the above conditions with different medications would cause **constipation**. This applies to the use of opioid analgesia and non steroidal anti-inflammatory drugs like ketoprofen in painful conditions, medications for the treatment of Parkinson's disease, mental health problems including depression, anxiety and psychosis.

In summary, the age related diseases and conditions cause **constipation** either directly, by causing immobility or as a side effect of the medication given for the treatment of the medical condition.

2 Immobility and sedentary lifestyle.

I have just mentioned this briefly above. Suffice it to say that immobility is an important cause of constipation in the elderly and this should be prevented as much as possible in order to deal with the **constipation.**

Sedentary life style is one of the most important causes of constipation. Exercise is like an engine to the body that stimulates physical and mental wellbeing. This includes the gastrointestinal tract by improving it motility and hence that of its contents which ultimately prevents constipation. As a result of several age related medical conditions that affect the elderly and general debility in the very elderly, mobility problems occur.

These are particularly common in fractured neck of femur, post operatively, in bed ridden patients, diseases of the spine, knees, legs, feet and osteoporosis with pathological fractures.

Senior citizens with Alzheimer's disease, Parkinson's disease, breathing and cardiovascular problems, including heart failure, severe ischemic heart disease, strokes and long term illnesses and conditions have little interest nor the ability in getting out of bed. In such cases **constipation** becomes a major problem.

Those who can be mobilized out of bed should be encouraged to do in order not only to prevent constipation but bed sores and vitamin D deficiency due to lack of sunlight.

3 Polypharmacy.

This is the administration of many drugs at the same time. This is quite common in the elderly and is usually due to either multiplicity of medical conditions that require different medications and sometimes to lack of review by health care professionals failing to cancel medications that are no longer needed.

We have seen in chapter seven that there are several medications that are responsible for constipation. Indeed the list in chapter seven is the tip of the iceberg; there are many more. The problem with polypharmacy is that if a senior citizen is on ten different medications, three or four of such drugs could be **constipating.** An example of this is when one is taking an Opioid medication like Codeine for pain but the symtom is not well controlled and Morphine is added. On top of these, the patient has mental health problems or Parkinson's disease and possibly also suffering from gout or arthritis and taking ketoprofen: the medications for these conditions cause **constipation** which means that it would be multiplied several times over.

The senior citizen may actually be bed-ridden due to these and other medical conditions such as diverticular disease, stroke, myocardiac infarction, post operative cancer with bed sores or hypothyroidism The potential for **constipation** in such cases would be immense. The effect of poly pharmacy here would actually be combined with immobility plus some medical conditions that cause constipation. In actual fact, this is not an imaginary scenario but the clinical reality of the problem of poly pharmacy with added effect of other causes of constipation.

First, there would be no need to give codeine if a stronger analgesia like morphine is also being administered which may be sufficient to relieve the pain of arthritis. Every attempt should be made to get the patient up for a few minutes/hours of walking with a frame or walking stick and some may surprisingly actually do some simple supervised exercises.

For these patients adherence to the 3FEV principles of fruits, fibre, plenty of fluids, some exercises and vegetables become even more important. These are the "3 female examples of virtue" in chapter one which in the elderly is a sine qua non if **constipation** is to be prevented or dealt with.

4. Cancer -particularly bowel cancer

The incidence and prevalence of nearly all forms of cancer tend to increase with age. This is no great surprise: often it takes a long time for carcinogens (substances that cause cancer) to reach the critical stage to cause changes in the cell and cause cancer. Someone who has been smoking for one or two years is far less likely to develop cancer of the lung than one who have been doing so for 20 - 35 years.

Besides there are certain premalignant conditions like intestinal polyps that are far more likely to develop into cancer after many years. Ofcourse, even young people can still get a fast growing polyp develop into cancer. The same is true of carcinogens in food: prolonged contact over the years with the mucosa of the colon is more likely to result in cancer than recent contact.

There is little doubt that older people are more likely to develop colorectal cancer, prostate cancer, uterine or bladder cancer.

Colorectal cancer causes obstruction directly leading to either a change in bowel habit with **constipation** alternating with diarrhoea or in some cases obstructing lesion with **constipation** as the main predominant feature.

The other forms of cancer mentioned above, prostate, uterine and bladder are more likely to obstruct the colon by direct extension and cause constipation. These cancers are still more common in the elderly than young people.

5 High risk of falls.

The elderly are particularly prone to falls and to be hospitalised for fractures. Fractured neck of femur, pelvis, shaft of femur, knees, spine, and colles fractures etc are more likely in osteopenic bones (due to osteoporosis) than normal healthy bones. Osteoporosis is the main catalyst for pathological fractures in the elderly.

Hospitalisation, operation and immobilisation are some of the most potent ingredients in the causation of **constipation**. Hospitalisation and immobilisation are very restrictive in terms of ability to get out of bed, walk or exercise.

Apart from fractures, certain diseases and conditions which predispose to falls are more prevalent in the elderly. Parkinson's disease, dementia including Alzheimer's, strokes, osteoporosis, arthritis, neurological diseases, poor eyesight and hearing all predispose to falls and accidents and confinement to bed and the development of **constipation**. These are all age related conditions.

In conclusion the vulnerability of the elderly to falls, certain conditions that predispose to falls, age related conditions and diseases, polypharmacy are the precursors or causes of **constipation.** These can to a large extent be prevented by paying attention to details and foresee these conditions and adopt appropriate measures to mitigate them and thus prevent **constipation.**

It is also most important to emphasise that despite the impediments, senior citizens are never too old to engage in sports and exercise to the limit of their abilities just like the youths and by so doing enjoy the same beneficial effects that exercise bestows. It may require a little more effort and determination but it is worth it in the end.

Lastly, in the words of **Albert Einstein**, "Life is like riding a bicycle. To keep your balance you must keep moving".

This applies to everyone but particularly to senior citizens.

CHAPTER NINE

Prevention of Constipation by Adequate Use of Diet, Hydration and Exercise

Food has always been an important part of our lives. Survival depends on it. What we eat can largely determine what and who we are. If we don't eat the right types of food rich in minerals, vitamins, proteins, carbohydrates, vegetables, fruits, fibre and some essential fat, we suffer the consequences. There are diseases and medical conditions that are linked to deficiencies of nutrients that are essential for proper functioning and development of vital organs and metabolism.

Fortunately much of the food we eat contains some of the essential nutrients that the body needs without our making any conscious effort: but that is not always the case. There are several food items that we eat that may actually do more harm than good. They may be rich in saturated fat like animal fat and not in unsaturated fat and lack the 3F &V.

There may also be too much carbohydrates in our diet and little or no fruits, fibre minerals or vegetables. A balanced diet is one that incorporates the essential nutrients above. They are sometimes not the most palatable and we would often prefer to satisfy our taste buds than our system requirements. We tend to go for what we are used to that tastes nice and palatable and easily obtainable without too much fuss especially when it doesn't entail hours of preparation and cooking. They may be easily obtainable or ready made with no consideration for food value or a balanced diet. Price is another deterrent: but buying at the right places is now common place due to intense competition by supermarkets, other outlets and greengrocers.

With modern cooking methods and in expert hands, rich nutritious food can be made quite palatable without sacrificing or compromising taste.

Our fore parents survived on a diet of vegetables, fruits, fish and hunted animals thousands of years ago. They were apparently quite satisfied with their lives then. They knew little about what we now know as essential nutrients.

No records are available about the prevalence of **constipation** or otherwise but anecdotal indicators point to fewer problems with constipation. Plenty of fruits and vegetables formed their staple diet. Fishing was one of their most cherished occupations as well as hunting. The former provided the essential oil rich in vitamins, protein and fish oil rich in unsaturated fat.

We also do know that our fore parents did not live a sedentary life style. They were usually quite active either as famers, fishermen, traders, craftsmen, shepherds, hunters or herdsmen. There were no white collar jobs and so little opportunity to sit in one place for prolonged periods of time which is an important factor in **constipation.**

Obviously we cannot go back in time to the dark ages but we can do things that promote mobility and activity. What about riding a bicycle to work, the corner shop, visiting friends or to the supermarket or better still walking at least some of the time and jogging? These are subtle forms of exercise which enhance activity, improve cardiovascular system and fight against strokes, hypertension, obesity, cancer, diabetes and, of course, **constipation.**

The most important factors involved in the prevention of constipation are the 3FEV (Fruits, Fibre, Fluids, Exercise and Vegetables) which have been mentioned in chapter one and some other parts of this book.

In this chapter I shall try to delve a bit more into each of these factors and throw some light on how they are likely to help in the prevention of constipation. We shall deal with each of these factors as follows: -

1. Fruits
2. Fibre
3. Fluids
4. Exercise
5. Vegetables

1. FRUITS

Fruits are important components of a balanced diet and play a significant role in the fight against **constipation.** The role of fruits, fibre and vegetables are similar but not necessarily identical and I have decided to treat them separately because of their uniqueness and individuality.

Fruits, and vegetables are good sources of fibre and there are also cereals that provide fibre but basically lack some of the other properties of fruit and vegetables, hence the uniqueness of each group despite similar functions in many respects.

At least 5 portions of fruit and vegetables are recommended daily. Some expert in the field go much further to 5-10 portions. They have important roles in the fight against not only **constipation** but heart disease, strokes, hypertension, diabetes, obesity and cancer.

What is a portion of fruit and vegetables? This according to latest information can be defined as follows: -

1. One portion is one large fruit such as pear, apple, banana, large slice of melon or pineapple.
2. One portion is also 2 smaller fruits such as plums
3. One portion is also one cup of small fruits such as grapes, strawberries, raspberries or cherries.
4. One portion is also one teaspoon of dried fruit.
5. A normal portion of vegetable (about 2 teaspoons)
6. One dessert bowl of salad.

Fruits contain fibre which as roughage provides bulk which helps in the propulsion of colonic contents and thus avoid **constipation.**

Fibre is that part of the plant where ordinary gastrointestinal enzymes cannot digest. This property is important as the bulkiness especially when there is enough fluid in the gut to ease the movement of contents through it. Some people appear to require more fluid to facilitate this process more than others. On average we require 6-8 glasses of water daily. This rises markedly in hot weather conditions especially in the summer when the temperature may be several degrees above normal.

It is important to eat certain fruits like apple, pear etc with the skin which provides insoluble fibre. This property is particularly important as there is no water to absorb from it and the bulk is present in the colon without water extraction which is an important property in preventing **constipation.** Water absorption from colonic contents results in hard faeces which is a potent factor in the genesis of constipation. There is also the real possibility that this insoluble fibre with its unique properties may help in stimulation of the colon thus increasing greater propensity for motility and peristalsis which militate **against constipation**. Secondly it is thought that the bulk and softening effect of insoluble fibre help to reduce intracolonic pressure which prevents diverticulosis and diverticular disease (small protrusions in the wall of the colon which fill with faecal material which becomes stagnant and inflamed).

The fleshy parts of fruits are rich in soluble fibre, though they mostly resist digestive enzymes, they absorb water in their passage through the gastrointestine tract and being rich in fibre they help to sustain the bulk in the colon which aids the passage of colonic contents.

There are several fruits with good fibre contents and, of course, other food benefits mentioned earlier. These are: -

Pawpaw (also called papaya)
Mangoes (+ unsweetened juice)
Apples (+ unsweetened juice)
Pear
Pineapples (+ unsweetened juice
Melon
Oranges (+ unsweetened juice)
Lemons
Lime
Grapefruit (+ unsweetened juice)
Cherries
Grapes
Raspberries
Strawberries
Banana. (high in calories if consumed in large quantities)
Plum
Prunes

These are only a few common fruits. There are several others that do the same job in providing fibre that aid in the prevention of **constipation.** Don't forget the saying that "an apple a day keeps the doctor away": but it also keeps **constipation** away!

2 FIBRE

Fibre is an important constituent of our diet because of some of its properties especially in the prevention of constipation. This has already been alluded to whilst discussing fruits above which are also good sources of fibre.

Apart from fibre in fruits and vegetables there is another type of fibre which are essentially cereal usually eaten at breakfast or cooked meals.

There are two types of fibre: soluble and insoluble

Soluble Fibre are plant food of which fruits and vegetable are the main sources. In many cases some plant food especially some fruits contain both soluble and insoluble types of fibre. For example apples and pears, provide both types of fibre. The skin is insoluble fibre but the juicy pulp fruit inside is the soluble fibre.

Soluble fibre is water soluble as the name implies whereas insoluble fibre is not water soluble. Soluble fibre unfortunately tends to slow down the movement of intestinal contents. This is only relative to insoluble fibre.

Insoluble Fibre does not dissolve in water. It provides bulk and is metabolically inert. It absorbs water as it transits the gastrointestinal tract rather than give up water for absorption which it naturally lacks. As a result of their bulking property, insoluble fibre tends to accelerate intestinal contents which accounts for its better **constipation-** preventing property.

In reality both soluble and insoluble fibres in essence help to prevent **constipation** and both still provide bulk but to varying degrees.

Apart from insoluble fibre from fresh plant - fruit and vegetables, there are other types of insoluble fibres which are more often used as breakfast cereals and others which are wholemeal food that require cooking and turning into a meal. There are usually whole grains meals which have undergone some treatment and parkaging. These are some examples: -

All-Bran
Bran Flakes
Weetabix
Shredded wheat
Muesli
Oat
Crushed Wheat
Whole meal bread
Whole wheat bread
Others -non cereals are: -
Brown Rice
Whole meal pasta
Whole wheat pasta
Whole wheat Spaghetti
Whole wheat Macaroni

3. FLUIDS

Fluid is an important constituent of the 3FEV-(the 3 female examples of virtue). Without adequate intake of water or other fluids, food cannot dissolve, fibre would be of little use because stool would be too hard and transit through the gastrointestinal tract would be greatly impeded and virtually impossible. Severe **constipation** would ensure, indeed would fall within the realms of **obstipation.** Fluid is therefore an essential part of the digestion, absorption, assimilation and the transportation of intestinal contents and non absorbed residue through the gastrointestinal tract down to the rectum and anus for evacuation. Intercellular fluid is essential in cellular metabolism.

We have already talked about, the reabsorption of water in the colon and how speed influences either too little absortion as in diarrhoea when the transit time is very short and in **constipation** when there is colonic delay leading to prolonged transit time thus allowing more water absorption in the colon and the production of hard stool resulting in **constipation**.

Although our main focus about water is its relationship to constipation, suffice it to say that we cannot completely ignore the role of water in general body processes.

About 60-75 percent of our lean body mass is water in the average 70 kg man. Variations are dependent mainly on the amount of fat of the individual. Fat has a low water content and thus reduces the total water per tissue mass.

The cells, contain about two thirds of the total water in the body and the rest are outside the cells and called extra cellular water.

Body water is derived maily from the food we eat and also from the fluids we drink. To maintain hydration we need to drink about 6-8 glasses of water a day though this requirement is higher in hot climates.

There are several uses of water in the body. I shall mention a few here. Water is essential for the elimination of waste such as urea from the body via the kidneys, helps to maintain temperature of the body: by sweating in hot weather and the evaporation of the sweat helps to cool the body and maintain normal temperature. Chemical reactions in the body need water. The life of body cells depend on adequate supply of water and so is the transport of nutrients in the body. Water also acts as lubricant around joints, has a protective influence as some sort of shock absorber for the brain, spinal cord (cerebrospinal fluid CSF), eyes (aqueous humor) and the foetus via amniotic fluid. Lastly, we cannot forget water in relation to the theme of this book: water aids digestion, lubricates colonic contents and thus eases the movement to its evacuation at the anus. Without water the stool is hard and the journey is slowed down resulting in a vicious circle where more water is absorbed by the gut and results in SHIID (**S**training, **H**ard stool, **I**nfrequency, **I**nsufficiency of stool and **D**ifficulty)

4. EXERCISE

"Exercise adds years to our lives and lives to our years". It is the engine that propels the force behind the fight against **coronary vascular disease, hypertension, strokes, cancer, diabetes, obesity and constipation.** It can lead the battle but cannot win the war without its comrades-fruits, fibre, fluids and vegetables which together make the 3FEV. We cannot, of course, forget the third element, CISSA, in this important orchestra. That is when the war can be won.

The benefits of exercise have already been alluded to in different parts of this book. There is little doubt that exercise provides the dynamics that aid and accelerate the proper functioning of our body systems. In the gastrointestinal tract it enhances the force behind gut motility- the basis of peristalsis that propels colonic contents to its final destination.

A regular exercise of about half an hour a day 3-5 days a week would go a long way to satisfying the above ideals.

There is, of course, a special exercise to beat constipation there and then if properly performed that is introduced in the last chapter of this book

Many people who exercise have different ways of doing so. Some enlist in the Gym. Others use video workouts, some have exercise equipment at home, others exercise by running or walking, skipping, riding a bike or walking to work or corner shop instead of using the car. There are various means. The ultimate aim and results are virtually the same provided there is commitment and some regularity.

5. VEGETABLES

Vegetables like fruit, fight constipation due to their fibre content and like it, their fibre contents have soluble and insoluble parts.

Unlike fruits which are often eaten raw, most vegetables are parts of a cooked meal except salads of course.

The fibre in vegetables play the same role in the prevention of constipation as for fruits and fibre per se already discussed above.

Unlike fruits some vegetables are "leafy vegetables" and provide a good source of insoluble fibre and some soluble as well.

The commonly used vegetables are the following: -

Peas
Green Beans
Cabbage
Brussel Sprouts
Broccoli
Okra
Lettuce
Cucumber
Cauliflower
Celery
Spinach
Asparagus
Green Peppers
Onions
Kale
Carrots
Sweet Potatoes
Beans

The list is endless because different parts of the world have different and peculiar vegetables and different species abound and some are eaten mostly in times of farmine or food scarcity.

Lastly we have seen the benefits of the **3FEV** in the fight not only against **constipation** but Diabetes, Ischaemic heart disease, Strokes, Cancer, Hypertension and Obesity. (**DISCHO**). We have noted how the important health and life style factors-Cigarettes, Salt, Sugar and Alcohol (**CISSA**) are important in the causation of not only constipation but **DISCHO**. Understanding the relationship between these factors and the health message behind them are crucial in the prevention of these very important conditions. It is almost beyond comprehension and belief that understanding the genesis of constipation could unravel the relationship between it as well as these other important conditions and their prevention.

CHAPTER TEN

Weight Loss that is Sustainable and Permanent by Adhering to these Concepts and Practices

The methods and concepts in this book do not stop with constipation, diabetes, ischemic heart disease, strokes, cancer, hypertension and obesity. One of the by-products is the fight against obesity itself which is clearly embodied in the "O" in **DISCHO.** We cannot throw away or discard an important and life changing by-product such as this; we have to utilize it for all its benefits.

This important by- product is the fight against obesity which has life changing health implications and ramifications that reverberates globally in a world where obesity is now assuming "epidemic" proportions.

This fight begins with the proper utilization of 3FEV. It is simple and basic in principle but requires adherence and determination in practice to achieve the desired goal of not only preventing constipation and DISCHO but also obesity and overweight We have seen in previous chapters the value of a balanced diet that is rich in fruits, fibre, plenty of fluids, good exercise routine and vegetables (The 3 Female Examples of Virtue). We discussed this in relation to the prevention of constipation initially and later in combination with CISSA to fight not only constipation but DISCHO.

We now have the opportunity to utilise these concepts in actual practice in the fight against overweight and obesity. I am sure many of my readers already know much about the Body Mass Index (BMI) as well as the normal range, overweight and obesity. It is also important to say briefly how BMI is measured and its significance

The Body Mass Index is a measure of our body weight that is based not only on actual weight but also on height. The important message here is that our weight should take into consideration how tall or short we are. This gives the net weight in relation to height. Two men 5 and 6 feet tall would each have quite different BMI with the taller man having the BMI or net weight advantage because of his height. The shorter individual would aim to reduce his weight more inorder to achieve the same BMI as the taller individual. Depending on the weight of the two people in this example, the taller gentleman could be classified as normal in weight or BMI whilst the short one could be either overweight or obese. In practical terms if you share the same weight as your friend but not the same height, you don't truly share the same weight or BMI. Your potentials for overweight, obesity and associated disease entities may therefore be completely different due to disparities in height.

BMI is calculated by weight in kg divided by height in metres squared. As a typical example if 4 friends each 80 kg in weight but are 1.58, 1.68, 1.84 and 2.09metres in height respectively, their BMI would be 32, 28.3, 23.6 and 18.3 respectively.

There are 4 BMI categories as follows-

Underweight =<18.5
Normal Weight = 18.5-24.9
Overweight=25-29.9
Obesity = 30 and above

In the example given above, the 4 friends would be Obese, Overweight, Normalweight and Underweight respectively despite each being 80 kg in nominal weight. The BMI gives their true weight relative to their heights. That is the weight that matters; BMI is measured in kg per metre squared and represents the actual weight that takes into consideration the height of each individual.

There are, of course, other parameters in determining weight and health such as the waist circumference but we shall not go into these in this book

People who are overweight or obese should idealy lose some weight. The weight loss programme is for adults who are fit with no serious medical problems. Those with health problems should first consult their General Practitioners before embarking on any weight reduction regime.

Under normal circumstances, men require about 2500 and women 2000 kcal daily to meet normal day to day activities. Unfortunately overweight and obese individuals exceed this limit. Inorder to get back to normal and ensure weight loss that is gradual, sustained and permanent, the NHS Choices recommends maximum 1900kcal for men and 1400 for women daily. A gradual weight loss of about 0.5-1kg (1-2 lb) per week is recommended. Where daily intake has been exceeded then any excess should be compensated for during the week by further calories reduction until the target is achieved. For example if a woman's calories intake for a particular day is 1800 kcal instead of 1400 then there is an excess of 400 which should be compensated for by further reduction during the week to meet the target. Weight reduction regimes are weekly and run for 12 weeks. The problem is what happens after the target weight has been achieved.

Unfortunately, most people go back to their usual habits and put back the weight which they had lost. Weight loss seems to be a temporary achievement of convenience which is often used to "look good" for an important occasion such as a wedding, interview, a date, a party, TV appearance or other engagements.

A sustainable and permanent weight loss can be achieved if there is a complete change in lifestyle and attitude so that weight loss becomes natural and not a struggle or an inconvenience that has to be practised and taught in a regimental fashion. When that is the case, weight loss ends when the circumstance that prompted it has been achieved.

Weight loss should be fun and the health message should be the dominant and determinant factor which supersedes every other consideration so that it becomes part and parcel of the individual's way of life and engrained in the psyche.

This can only be achieved if the simple messages of 3FEV and CISSA become a way of life. The messages incorporate among other things, adequate diet that is rich in fruits, fibre and vegetable as well as plenty of fluids and exercise that become part of everyday life. The CISSA element is concerned with breaking bad lifestyle or unhealthy habits and addictive behaviour particulary cessation of smoking, reduction in salt, sugar and alcohol intakes. The latter two are behaviours which promote excessive calorie intake in addition to other effects which do not impinge directly on weight.

The main thrust here is the adoption of an healthy lifestyle which automatically deals with any weight and brings down BMI. What is most interesting about the 3FEV is the fact that vegetables and fluid (water) are the least sources of calories but good sources of satiety.

Two glasses of water could fill the stomach so much that one forgets that he/she has not actually eaten any food with calories. Vegetables, especially leafy vegetables would almost do the same as they contain little or no calories. If these are combined with exercise that is regular and sustainable then weight is shed on a regular basis without much thought or attention. Let me emphasise here that I do not advocate not eating adequate calories but reducing the amount of calories in overweight and obese individuals only.

We shall now look at a few examples of calorie contents of some food items compared to common vegetables according to NHS choices-

Weight.	Food Item	Kcal	Fat content
100 g	celery (boiled)	8	0
100g	Broccoli	24	1
100g	Cauliflower (Boiled or steamed)	28	1
100g	Asparagus (boiled or steamed)	26	0
100 g	Cabbage (Boiled or steamed)	14	0
100 g	Carrot	35	0
100g	Spinach	19	1
100g	Apple	49	
100g	Banana	95	
100g	Avocado	191	20
	(Other Food Items)		
100g	Weetabix	362	2
100g	Basmati Rice	357	3
100 g	Brown Rice	357	3
100g	Sweet Potatoes	115	0
100g	New Potatoes	75	0
100g	Chinese Egg Fried Rice Takeaway	678	18
100g	Long Grain Rice	303	4
125g	Skimmed Milk	10.5	0
125g	Semi Skimmed Milk	61	2
125g	Whole Milk	about 80	about 4.5

I have used a few examples of common food items to demonstrate the calories contents of food. What has become apparent is the virtual exponential rise in calories from vegetables to fruits and other food items. This is not entirely surprising as vegetables are mostly leafy with high fibre content and little or no calories. Fruits occupy the middle cadre and like vegetables are rich in fibre but also have the juicy succulent pulp which is bound to raise the sugar and calorie content I deliberately separated potatoes from other vegetables because of their supposedly high calories content but surprisingly to some people100g of new potatoes have less calorie content than 100g of bananas and only 26calories more than 100g of apples.

Milk is another area which has been shown above in which there is again a steep rise in calories from skimmed, semi skimmed to whole milk.

Other food item like rice has a high carbohydrate or sugar content and the calories are consequently high and in some cases quite high.

Despite the variations in calories, in order to have a balanced diet we would usually need to have at least some of the carbohydrate high calorie-laden food items. In other cases where there are other food items with lower calories content, they can be substituted. Potatoes, particularly new potatoes, serve that useful role of providing carbohydrate and low calories.

The most important point in losing weight is a change in food habit and good physical activities. Changing food habit is not always easy but it can easily be done in a gradual step by step fashion. It is also important to eat more vegetables and learn to make them more palatable. Vegetables in the 3 FEV should form the largest portion of our diet. From there one can proceed on to fruits which like vegetables have high fibre content but a little more calories. The high energy, high calorie food items like rice and some pasta should form the least portion in our diet. Consumption of fatty food items should be reduced.

We have also seen the exponential rise in calories from skimmed, semi skimmed to whole milk. Inorder to effect weight reduction, a gradual change from whole to semi skimmed and eventually to skimmed milk is highly recommended.

Infact these gradual, hardly noticeable changes in diet and life style would make the difference between losing weight, sustainability and permanently maintaining it or following a dieting regime; losing weight and putting it all back again after that. Without a complete change in diet and increased physical activities like exercises, maintenance of weight loss would only be an illusion, a mirage or a short term accomplishment which has to be repeated several times a year necessitated by cosmetic considerations only but not for health and wellbeing advantages. No weight loss programmes including so called "magic tablets" can ensure sustainability and permanency without long term considerations of calorie intake and regular exercise.

The maintenance, sustenance and permanency of weight loss should incorporate the vision and attitude of a long term strategy which requires commitment and determination initially but gradually easing and merging into a lifestyle that is automatic and second nature which becomes unnoticeable by the individual except as a normal way of life.

CHAPTER ELEVEN

Laxatives and Medical Management of Constipation

Laxatives are medicines used to treat constipation. They are in different forms-liquids, tablets, capsules, suppositories (for insertion into the rectum) and enemas.

We have already seen that bowel habit varies considerably between individuals with some opening their bowels 1-3 times a day and others 1-3 times a week and indeed, there are others who lie at the periphery of these so called normal. It is therefore important that everyone should know their usual bowel habit and seek medical help when there is a persistent change despite the 3FEV measures. A change in bowel habit requires a medical input from the family doctor.

Other factors which I have highlighted in different chapters in this book should also be considered especially the effect of currently used medication, change of environment and diet etc.

Laxatives are only prescribed when other measures have failed or constipation is considered to be so serious that it is likely that straining could worsen an already existing condition such as angina or haemorrhoids (piles- risk of excessive rectal bleeding).

Laxatives have also been used for the expulsion of parasites after treatment with antihelminthic medication (medication used to expel intestinal worms). They are also sometimes used for drug induced constipation, to clear the gastrointestinal tract prior to surgery and for radiological procedures.

Laxatives have serious side effects which limit their use especially when alternative methods of management of constipation have not been seriously considered and tried.

These side effects include: -

- Dependence which worsens constipation as increasing doses have to be administered which usually ends with loss of effectiveness and worsening of constipation.

- Intestinal obstruction which can occur with bulk forming laxatives without adequate fluid intake.

- Laxative abuse leading to hypokalaemia (low serum potassium) which can result in serious consequences especially due to effect on the heart and tissues.

- There are other minor side effects such as flatulence, abdominal distension, faecal impaction, abdominal cramps and hypersensitivity.

There are 6 groups of laxatives in use. These are: -

1. Bulk forming laxatives
2. Stimulant laxatives
3. Faecal softeners

4. Osmotic Laxatives.
5. Bowel cleansers
6. Peripheral Opiod receptor antagonists

I shall now briefly examine the mechanism of action of the different groups of laxatives and the medical conditions in which they are commonly used. They may also be associated factors which should be satified to prevent possible side effects which as observed above can occur with laxatives.

1 Bulk Forming Laxatives

Bulk forming laxatives or fibre supplements increase faecal bulk in the same way as fibre does and by so doing stimulate peristalsis. This "bulking" effect is potentiated by the absorption of water which helps it to swell and increase faecal bulk. This effect then stimulates the colon to squeeze stool towards the rectum and anus for evacuation.

Bulk forming laxatives should not be used unless there is a reason fibre can not be increased in the diet. They are therefore useful for those with small hard stools. The 3FEV should be used first.

It is also important to emphasise adequacy in fluid intake if using bulk forming laxative as otherwise it could cause intestinal obstruction.

They are often used in the management of patients with ileostomy, colostomy, anal fissure, haemorrhoids, diverticular disease, irritable bowel syndrome and as part of the management of ulcerative colitis.

Here are few examples of bulk forming laxatives-

- Ispaghula.
  In which the following are examples-
  Fibogel,
  Fibrelief
  Ispagel orange
  Regulan
  Isogel

- Methylcellulose which Celevac is a typical example
- Stercula typified by Normacol and Normacol Plus

2. Stimulant Laxatives.

Stimulant laxatives stimulate nerve in the colon which causes colonic muscles to increase motility and squeeze harder to propel stool for evacuation.

Its mode of action may lead to abdominal cramps and should not be used if there is any suggestion of intestinal obstruction. Excessive use may also cause diarrhoea.

The parasympathomimetics (medications that stimulate the parasympathetic nerve), bethanecol, distigmine, neostigmine belong to this group of stimulant laxative and stimulate parasympathetic gut activity and this increases intestinal motility and thus increased speed and movement of intestinal contents thereby preventing constipation. You would recall from previous chapters that increased parasympathetic activity is associated with increased colonic motility, peristalsis, propulsion of colonic contents and ease of evacuation. That is what these parasympathomimetics do.

However, there are hardly used for their gastrointestinal effect and if used at all, intestinal obstruction must first be excluded and should never be used after bowel anastomosis as breakdown of anastomosis could occur due to increased bowel motility.

These are some examples of stimulant laxatives-

- Bisacodyl-used preoperatively and before radiological procedures and is contraindicated in acute surgical abdominal conditions, severe dehydration and in acute inflammatory bowel disease.
- Dantron is a stimulant laxative used only to relieve constipation in terminally ill patients because of potential carcinogenic risk.

Exemplified by Co-danthramer and Co-anthracite.

- Docusate Sodium is used for constipation and for abdominal radiological procedures.
- Glycerol or Glycerine are suppositories which should be moistened in water before use.
- Senna typified by commonly prescribed stimulant laxative Senokot.
- Sodium Picosulfate typified by Dulcolax, commonly used for bowel evacuation before abdominal endoscopic and radiological procedures especially of the colon.

3 FAECAL SOFTENERS

Faecal softeners work by softening the stool, helps to wet it and facilitate easy passage

It is typified by Liquid Praraffin which is a lubricant that helps to lower resistance in the passage of stool to the anus for evacuation.

Liquid Paraffin is contraindicated in children under 3 years of age. It may cause irritation of the anus due to staining and seeping especially after repeated use.

- Arachis Oil enema is used to soften impacted faeces for evacuation but should not be used for children under 3 years of age.

4. OSMOTIC LAXATIVES.

They exert their action as the name implies by a process of osmosis by retaining fluid or by drawing fluid from the body into the colon.

This process increases the easy passage of colonic contents in the bowel down to the anus as a result of lowered resistance due to the lubricating effect of soft rather than hard stool. Please recall the features of constipation (SHIID) in chapter one

Here are some of the commonly used osmotic laxatives-

Lactulose-Is not absorbed from the gastrointestinal tract and produces its effect by osmotic action with its effect amounting to osmotic diarrhoea which in the face of constipation helps to clear any solid or hard faecal matter.

It may take up to 48 hours for its effect to manifest.

It is contraindicated in conditions such as intestinal obstruction and galactosaemia. It may cause abdominal discomfort, cramps and excessive wind.

Macrogols (Polyethylene Glycols) are polymers which are inert and exert their osmotic effect by sequestration of fluid into the colon; an action which may cause dehydration. For this reason, it is important to drink plenty of fluids whilst on this medication.

There are several preparations with specific names-

- Macrogols Oral Powder
- Laxido Oral Powder
- Movicol
- Movicol-Half
- Movicol Paediatric Plain

Other Osmotic laxatives in common use are-

Magnesium Salts in which the following preparations are known-

Magnesium Sulphate a common preparation in parts of Africa called Epsom Salt with sour/bitter taste. It has a rapid bowel evacuation action in about 2-4 hours, usually taken in a glass of water before breakfast.

It should be avoided in hepatic and renal diseases. There is some risk of Magesium accumulation.

- Magnesium Hydroxide
- Magnesium Hydroxide with Liquid Paraffin

There are also bowel Cleansing preparations which are useful in abdominal surgical procedures, endoscopy and radiological procedures.

These are few examples -

Phosphate (Rectal)
- Phosphate Enemas

Sodium Citrate (rectal) as-
- Microlax Micoenema
- Relaxit Mico-enema

Micolette Micro-enema

5. BOWEL CLEANSING PREPARATIONS.

Are not for the treatment of constipation but used to cleanse the bowel before procedures such as colonoscopy (looking into the large intestine using an endoscope called a colonoscope), surgery of the large intestine and for the removal of faecal solid material before radiological examination.

They usually require attention to the fluid requirements of the patient to maintain fluid and electrolyte balance especially in the elderly and children.

Examples include the following

-   Macrogols
-   Magnesium Citrate
-   Oral Phosphates
-   Sodium Picosulphate with Magnesium Citrate.

6.   PERIPHERAL OPIOID RECEPTOR ANTAGONISTS.

These are used in conjunction with other laxatives for the treatment of constipation due to opioid medication such as Codeine, Morphine, Co-Codamol, Pethidine etc.

Peripheral opioid receptor antagonist is exemplified by Methylnaltrexone commonly used for terminally ill patients experiencing constipation due to opioid analgesia.

The purpose of this chapter is to highlight and give examples of commonly used laxatives and their modes of action.

The general and possible side effects have also been stressed in some of the laxative medications.

The next chapter is very important in showing how the use of laxatives can be greatly minimized and avoided in several cases by the use of methods which will become clear in the next chapter.

That will only be the case if attention is paid to details without jumping to conclusions when the methods have not been fully tried and tested for several weeks.

It is also important to stress again that before using the methods, any change in bowel habit should be an indication for seeking medical attention.

CHAPTER TWELVE

Short Exercise, Positions Etc that Yield Immediate Results if Properly Carried Out

We have seen in preceding chapters that there are several things that we can do to prevent constipation. However despite these, there are still times when "normal" constipation appears to be resistant to all the measures. This chapter deals with other measures, some of which appear not have been fully considered and others that have been known but which no reasons or mechanism of how and why they work have been given.

However before going into these additional methods, I would like to emphasise again the fact that these measures are unlikely to work in isolation without the principle of 3FEV being applied until it becomes part of the overall norm. We still have to remember that **F**ruits, **F**ibre, **F**luids, **E**xercise and **V**egetables form the basis or cornerstone in preventing constipation and without them the measures in this chapter might not work. In a previous chapter, there have also been some emphasis on the quantities of the fibre intake and the fact that fruits, fibre and vegetables are basically fibre though with varying degrees of soluble and insoluble fibre contents.

Apart from the 3FEV, this chapter will now consider additional factors which are crucial in resistant cases of constipation where any underlying conditions have been excluded. This actually means that **a change in bowel habit** has been ruled out by a doctor. What is a change in bowel habit? This just means what it says but likely to be indicative when constipation has some of the following features-

- persistent prolonged intervals between normal times of bowel opening
- constipation alternating with diarrhoea.
- constipation with rectal bleeding
- constipation with blood and mucus in stool
- constipation with severe abdominal pain
- obstipation
- constipation that is unrelieved by usual measures like laxatives
- constipation in which all the possible causes have been excluded, and is prolonged and unresponsive to usual measures.
- constipation with abdominal distension

In this chapter we shall deal with "normal" constipation and look at how immediate results can be obtained and consider it under the following headings-

1. Gastrocolic Refex
2. Having breakfast before using the toilet
3. Early morning glass of water or juice
4. The Effect of Gravity on bowel movement
5. Positioning- the profound beneficial effect of squatting position
6. Use of footstool or SquattLooStool
7. Left leg flexing and positioning on toilet seat
8. How Spot Exercise could be the ultimate answer that yields immediate results.

I shall now consider each of these methods separately although some actually come from the same principle and will be discussed in relation to the overall principle that covers each of them.

1 How to make use of the Gastrocolic Reflex to combat constipation.

Gastro colic Reflex is one of the most forgotten and ignored phenomena in the fight against constipation. It is now time to remember that when other methods appear to have failed we should not hesitate to harness the gastro colic reflex.

What is the gastro colic reflex? The presence of food or drink in the stomach causes distension and stretching of the muscular wall of the stomach. This distension causes a sensation by stretch receptors in the wall of the stomach. There is a spread of this stretch which activates colonic motility, peristalsis and movevemet of colonic contents downwards towards the rectum.

If there is high intensity of this reflex, the contraction spreads rapidly from the asending to the transverse colon which triggers mass movement that propels the colonic contents to the descending, sigmoid colon and the rectum. It does appear that gastro colic reflex also depends to a large extent on the quantity of food and fluid in the stomach to stretch and engage enough muscular wall of the stomach to cause effective colonic reflex.

It is important that enough bulk reaches the rectum to cause effective pressure and sensation and the stretching of the rectum is sufficiently adequate to promote bowel emptying.

It also appears that the mechanoreceptors in the muscle wall of the intestine initiate the reflex contraction of the sigmoid colon and there is relaxation of the internal anal sphincter to accommodate and allow the passage of stool into the anal canal. However, there is contraction of the external sphincter of the anal canal in which there is some conscious control which otherwise would result in incontinence and soiling when we are not ready to empty the bowel particularly where circumstances are not right. This could be when there is actually no toilet in the vicinity and we have had a meal which has initiated a gastro colic reflex that has created the urge to defaecate but without the right circumstances. That is where conscious restraint and delay come into play.

Defaecation will not then occur until there is a conscious effort to relax the external anal sphincter to allow bowel emptying at the appropriate time and place. The degree of this control varies with individuals and the amount of bulk and fluid presented to the sigmoid colon, rectum and the anus. Sometimes the urge becomes overwhelming due to such factors and accident occurs.

Not only the external anal sphincter but also the puborectalis muscles need to be relaxed to allow defaecation to occur.

It also appears that under these circumstances, another often ignored factor, gravity and position, come into play to forestall accidents. It does appear that in the face of overwhelming urge the sitting position with the legs partly or fully extended and brought together would better resist accidents than upright posture or flexion of the thigh and parting them which would tend to relax the external anal sphincter and puborectalis muscle and also straighten out the ano rectal angle into a straight line.

I shall come to the question of correct posture in a positive sense later, however, suffice it to say that conscious and voluntary relaxation of the external anal sphincter and the puborectalis muscle and the adoption of the correct posture and position aid and make it easy for bowel evacuation. The reverse is also the case in trying to prevent incontinence as seen above.

It looks very likely that the gastro colic reflex is governed and controlled by an higher cephalic centre in the brain with overall conscious control: otherwise we would empty the bowel after every meal or drink but that is not the case. It therefore means that the gastro colic reflex only becomes operational in times of need. This need would occur if there is a constipating episode with bulk and distended colonic contents especially in the recto sigmoid region. It appears that this reflex is more likely to occur if there is sufficient distension of the stomach with food or drink sufficient to stretch the gastric muscle fibres to a certain critical level which impacts on the nerve endings with sufficient engagement of gastric muscle fibres to ensure transmission to the colonic neural mechanism to effect this reflex. This is analogous to Sterling's law of the heart which in same fashion applies to the stomach ("The greater the stretch of cardiac muscles, the greater the force of contraction. This means that when there is an unusual increase in volume of blood entering the heart, the ventricular wall stretches causing the cardiac muscles to contract more forcefully") This principle would then apply to the gastro colic reflex requiring sufficient engagement of gastric muscle fibres and mechanoreceptors and the spread of impulses down to the colon to initiate this reflex.

The next two topics- having breakfast before using the toilet in the morning and/ or water or juice are based on empirical evidence underpinned by the gastro colic reflex just described.

2 Having Breakfast Before Using the toilet.

For people who regularly experience constipation despite adopting the right measures including 3FEV and when a change in bowel habit has been excluded, having breakfast before going to the toilet can be quite useful.

This utilises the gastro colic reflex which we have just talked about above. The presence of food in the stomach is likely to initiate this reflex.

Initiation of the reflex causes stretching of the colonic wall which descends downwards to the sigmoid colon, rectum and anus. Under the right circumstances, the urge to go to the toilet becomes intense.

The gastro colic reflex which can take as early as 15 minutes to 2 hours is the logical explanation for initiation of this urge to use the toilet. It is unlikely to be the breakfast that passes to the rectum within such a short period of time. It would take much longer for digestion to take place and the residue to traverse almost a 9 metre length, undergo all the changes, with more absorption of nutrients and water in the colon. It would be quite a long journey. Fortunately that is not what happens: the stool is already present in the recto sigmoid colon. It is the gastro colic reflex that helps it reach its final destination.

3 Early Morning Glass Of Water Or Juice.

It has long been known that having a glass of water, orange, pineapple, apple juice (or breakfast above) etc accelerate toilet use in the morning.

This again is the product of gastro colic reflex as the presence of food or good quantity of fluid or juice in the stomach stimulates this reflex.

There may be an additional factor here, particularly with fruit juice, which may actually make the journey, help to wet and increase bulk of the colonic contents especially fibre already present in the gut and also helps to lubricate and provide a smooth non-friction background for easy movability of bulk to its destination. The possible direct involvement of juice would of course take a longer time in normal subjects except in circumstances where the subject is already prone to shortened transit time as in diarrhoea.

It does appear that under circumstances where gastro colic reflex is harnessed for the purpose of relieving constipation, a good quantity of fruit juice or water could make the difference- perhaps up to two or three glasses in bad cases in certain individuals with history of easy susceptibility to constipation when other methods have failed.

When this is combined with the special exercise and position to be described shortly, it could make a world of difference.

4 The Effect Of Gravity On Bowel Movement.

RECTUM AND ANAL CANAL

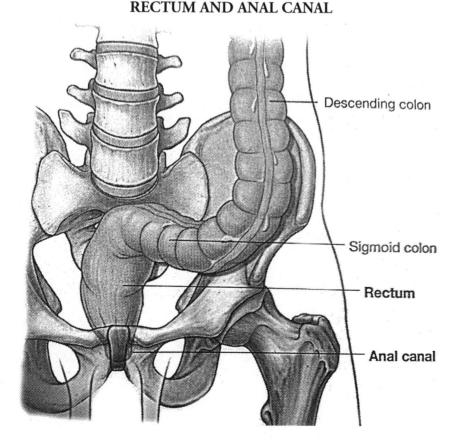

With Kind Permission of ELSEVIER Churchill Livingstone.

This is an area that has not been fully explored that gravity could have some effect on bowel movement.

It is well known that people who do sedentary jobs are more likely to have constipation than those who are on their feet most of the time whether the job involves some physical activity or not.

Bed-ridden patients are more prone to constipation than those who are mobile even if that involves just standing up without any physical activity.

Post operative patients who are not mobile suffer more constipation than those who are mobile or even just able to stand.

Patients who attempt to open their bowel lying down find it more difficult than those who are mobile and able to use the commode or chamber pot.

People with constipation tend to benefit more from exercise which involves standing or sitting than lying down exercises. The spot exercise which I wish to introduce later in this chapter is carried out in the standing position.

The rationale of these observations is that colonic contents gravitate downwards and that is better achieved if the force acting on it is in a downward direction, and especially, more so in the descending and recto sigmoid colon in the presence of bulk and colonic motility.

The effect of this theory and exposition is that we should now begin to consider constipation also in terms of the position we assume all day. Laziness breeds constipation: usually someone who spends most of the time in bed is more likely to complain of constipation than one who is up and doing.

The effect of gravity as an aid to bowel emptying and reduction in constipation is based on the above observations and many others. The evidence is therefore largely empirical (**"of medical treatment based on practical experience rather than scientific proof"**) though few people would deny the fact that lying down in bed all day is not going to make your bowels equally "lazy" and getting up and about even without exercising adds a stimulating influence to the body and bowel action.

The next section will show that when the recto-sigmoid-anal angle is obliterated and recto-anal angle straightens out, then bowel evacuation becomes unhindered The effect of gravity is maximal when acting in a straight line. This benefit can be mobilised and utilised by employing the methods in the next sections.

The evidence of the effect of gravity in easing constipation and facilitating the urge to defaecate are both anecdotal but essentially and mostly empirical.

The next three sections, namely: the effect of squatting, using the footstool and having the left foot on the toilet seat are all derived from the same principle of the straightening of the ano rectal angle in the squatting position as opposed to the sitting position. There are strong pieces of empirical evidence of better evacuation in these positions.

5. **Postioning-The Profound Beneficial Effect of the Squatting Position.**

RECTUM AND ANAL CANAL

With Kind Permission of ELSEVIER Churchill Livingstone.

There has been the evolution of the modern toilet as we now know it over several years. In the course of this evolution, there have been major changes. Our fore parents knew nothing of the toilet with flush system and having to sit down to use the toilet..

In England, the precursor to the modern flush toilet system was designed by John Harington in 1596 and was not widely used till the late 19 century. Thomas Crapper became one of the first early makers of toilets in England but not the inventor.

The widespread use of the precursor of the modern flush toilet in England was in the late 19th and early 20th centuries. By then more people became introduced to what has now become the toilet in England.

Before then, pit toilets were the norm and several names were used to describe them including latrine, lavatory, Privy, loo, the netty, the Jack (Irish) and WC. An improvement on the pit toilet was the chamber pot or commode. Many of these names are still used today in different regions and countries.

The important fact to note is that in pre Victorian and most of Victorian times, these toilets had no sitting facilities. Squatting was the norm.

Toilet facilities took place outdoors in outhouses or in latrines with no flush systems. The pail closet became introduced in England and France later to help reduce sewage and sanitation problems in fast expanding towns and cities. In the early days of the introduction of "these modern facilities" only few people could afford them.

Our civilization and modernisation of the current toilet system to incorporate a seat and the abandonment of the squatting position has not been without some disadvantages in the ease, completeness of evacuation and the fight against **constipation**.

In the Indian subcontinent and parts of Asia, the squatting position is still quite popular and toilets seats are made to allow both squatting and sitting and satisfy cultural sanitary requirements of washing after toilet use.

The anal canal and the rectum form an angle of about 80 degrees. This angle is maintained by the action of the puborectalis muscle. In the squatting position during the act of defaecation, there is flexion of the thighs and this straightens out this angle and facilitates the evacuation of rectal contents with much greater ease.

In other words, when we open the bowel in the squatting position, the rectum and the anal canal are in a straight line. The abdominal pressure is raised, the pelvic floor is lowered and the relaxation of the external sphincter allows stool to pass through the anal canal much easily.

I shall make a comparison here to driving a car through a road bent at 80 degrees and a straight road devoid of any bend. Obviously, the driver has to slow down considerably in the former and maintain a good speed in the latter. The same principle applies to sitting down to use the toilet compared to squatting. Despite this apparent handicap, there are a few things we can do to enjoy the benefits of both worlds without compromising the comfort of sitting down.

We do not want to go back to prehistoric, medieval or even pre Victorian and Victorian times and use the squatting position again. We can "kill two birds with the same stone" - enjoy our modern sitting position as well as obliteration of the 80 degrees angle between the rectum and anus to achieve a straight line. This would reduce many cases of **constipation** when combined with the other factors.

The next two topics explore and explain how this can be achieved easily and comfortably without compromising the comfort of sitting down. The evidence for these are empirical.

6.   Use of the Footstool or the SquattLooStool

There is little doubt that the squatting position which our ancestors used for several generations and which stood the test time for so long had undeniable advantage; the ability to straighten the anorectal angle and facilitate bowel motion and sometimes even in the face of **constipation** which would present great difficulty with our current sitting method.

Unfortunately the primitive practice of squatting also had a disadvantage; namely, it is very tiring and uncomfortable for those who still use it.

Our ancestors and fore parents were able to endure such problems that would be quite intolerable today. Certainly we are not going back to the dark ages.

We can enjoy the advantage that the squatting position imparts and still have that measure of comfort of sitting down which we enjoy and often take for granted today.

Raising the leg with a footstool or one of the specialised ones like SquattLooStool achieves the same objective and best of two worlds without compromising comfort.

Using a footstool flexes the thigh at the hip and which virtually abolishes the anorectal angle which ensures better evacuation in the same manner as squatting but without the inconvenience and tiredness of the legs in squatting.

7. Flexing the Left leg and Positioning on The Toilet Seat.

There is another method which appears to have escaped researchers and thinkers which I wish to bring to their attention and that method involves using the left foot only to achieve the same goal as the footstool and other devices to raise and flex the thigh.

Cost aside, there are people who would rather not bother taking something like the footstool or the specialised derivatives into the toilet room.

For those who may not want the little inconvenience of a footstool, the answer lies in the use of the left foot.

This method makes use of the anatomy of the colon and the rectum. The caecum and ascending colon are on the right side of the abdomen whilst the transverse colon connects it to the descending colon and sigmoid colon to the left. Although the rectum and the anal canal come to assume a more central position they are essentially anatomically connected to the left structures.

Putting the left foot on the toilet seat, flexes the left thigh at the hip and obviates the need to use the footstool. In this position the angle between the right and left thighs should be about 100-140 degrees to achieve the same advantage as squatting or using the foot stool. The recto-anal angle should open and straighten in the same manner as described above. If other factors already described are satisfied bowel emptying should proceed with greater ease and some cases of **constipation** adequately dealt with. This method is based on strong empirical evidence. I must say that full research on such intimate matters are not easy to conduct.

8. How Spot Exercise could be the Ultimate Answer that yields Immediate result.

We have already noted that "exercise adds years to your life and life to your years" but this special exercise could also take away **constipation** from those years if properly performed and the other factors already outlined are satisfied.

This is a specialized exercise that crystallizes the value and benefit of physical activity in the fight against constipation.

The importance of exercise generally has already been alluded to in the prevention of diabetes, ischemic heart disease, strokes, cancer hypertension. and obesity (see DISCHO in chapter 8). Exercise is part of the 3FEV which is vital in the prevention DISCHO and also **constipation**.

In chapter eight we also saw that cessation of smoking (**CIgarettes**), reduction in **S**alt, **S**ugar and **A**lcohol intake (**CISSA**) affect our general health particularly the prevention of DISCHO. The relationship of these factors to constipation have already been discussed.

Before I discuss this specific exercise, it is important to state that it should not be carried out in isolation but in tandem with the factors discuseed in this chapter especially, the mobilization of the beneficial effect of the gastro colic reflex like having 1-3 glasses of water or juice in the morning on waking up or breakfast several minutes before using the toilet.

This exercise recognises the beneficial effect of general strength and fitness imparted to all parts and organs of the body by exercise. That beneficial effect is equally true of the colon where there is the possibility of increase in its motility aided by the effect of gravity as this exercise is carried out in the standing position.

Though the exercise is not particularly strenuous, the elderly, the sick and those with cardiovascular problems should first consult their family doctor before embarking on this simple exercise which should take about 20-50 minutes. It does not have to be continuous but interspaced with periods of rest. Like anything else in life, "practice makes perfect".

The exercise should be carried out not at odd times but during the individual's normal toilet time. Some people should have immediate results but for others it could take weeks of practice to perfect the art of this simple but special exercise.

It works better by wearing a pair of slippers with soft/firm slightly raised heel and consists of alternately raising the heels and lowering in a fast rhythmic fashion and when possible flexing and extending the forearms at the elbow. Raising the heel starts from the mid foot but keeping the forefoot steady on the ground. Many people may find it more convenient and easier to hold onto something rather than flex and extend the elbow which could be tiring and others may find that they don't have to hold onto anything. You should carry it out in your own pace especially initially until the routine is mastered. The first sign of success may be heralded by flatus. The urge to use the toilet could become intense in the next few minutes. As already stated above, the elderly and people with heart disease and similar medical problems should consult their doctor before embarking on this exercise. It's also important to look at all exercise routines as graded which means that the duration and intensity should be increased gradually over time especially for those who do not exercise regularly. Some people may not obtain immediate results until they have mastered the routine, are able to exercise at a good fast pace and above all combine the exercise with other measures already described in this book. People who fail to achieve any results despite a combination of these measures should preferably consult a doctor.

Thus this short exercise is a combination of adding vigour and drive to a sluggish or lazy bowel aided by the beneficial effect of nature- gravity. Other factors 3FEV, food and drink to stimulate gastro colic reflex and position on the toilet seat should be combined with spot exercise.

We have learnt quite a lot in this book about constipation, the causes, prevention and the treatment of this often forgotten inconvenience. We have also noted that sometimes it is much more than just an incovenience but that there may be an underlying medical condition that requires a diagnosis and treatment. I have pointed out the possible symptoms when a doctor should be consulted. I do not expect the lay reader to digest this book from cover to cover as some sections require some background medical knowledge but the lay reader can still use such sections as a reference guide, particularly when to consult a doctor.

Perhaps one of the most striking element in this book is the knowledge that the lack of factors (3FEV) responsible for the causation of constipation are also the factors required to fight diabetes, ischemic heart disease, strokes, cancer, hypertension and obesity (DISCHO) This relationship is very important because individuals who are healthy, exercising, eating adequate fruits, vegetables including fibre and drinking enough fluids are far less likely to suffer from constipation and DISCHO.

There are also additional factors in the fight against ill health. These factors also bear almost a direct relationship with the avoidance of constipation because they are directly or indirectly related. These factors are- the cessation of Cigarette's smoking, reduction in salt, sugar and alcohol intakes (CISSA) Tobacco smoking is one of the most important factors in the prevention of coronary heart disease, cardiovascular diseases in general, pulmonary diseases and cancers, all of which cause ill health or have direct relationship with constipation. Salt, sugar and excessive consumption of alcohol are important in the causation of cardiovascular diseases- hypertension, strokes, cononary heart disease, also diabetes and obesity. As shown above these diseases are also directly or indirectly related to constipation. There is therefore a very important relationship between **CONSTIPATION, 3FEV, CISSA** and **DISCHO.**

We have seen how the benefits of 3FEV and the CISSA message are critical in the fight against DISCHO and OBESITY and how to mobilise these to effect **weight loss** that is sustainable and permanent through low calorie diet and exercise.

In this book we have looked at constipation not as an isolated problem but also the relationship with other disease entities and the interplay between them and the global nature of the problems that link them together and how to deal with these problems in general and constipation in particular. The main message is one of prevention. Lastly CAESAR was a powerful Roman Conqueror-not easy to defeat, so is CISSA.

REFERENCES

1. British National Formulary; (ed. 68) BMJ Group and the Royal Pharmaceutical Society of Great Britain (c) 2014

2. Cuschieri, A. Giles, G.R. Moossa, A.R. (1982) Essential Surgical Practice. Bristol London Boston 958-1010: Wright BSG.

3. Drake, R.L. Vogl, W. Mitchell, A. (2005) Anatomy for Students. Philadelphia. Edinburgh. London. New York. Oxford. St. Louis. Sydney. Toronto: Elsevier Churchill Livingstone.

4. Boon. N. A. Colledge, N.R. Walker, B.R. Hunter, J.A.A. (2006). Davidson's Principles & Practice of Medicine Edinburgh London New York Philadelphia St Louis Sydney Toronto: Churchill Livingstone Elsevier.

5. Dirckx, J.H. (1997) Stedman's Concise Medical & Allied Health Dictionary. Baltimore: Williams & Wilkins.

6. New England Journal of Medicine 1998; 339: 1100-1104

7. Peadiatric Clinics Of North America 1954; 4: 940-962

8. Green, A (November 1998) Dr. Green.Com. Milk and Constipation.

9. Hardings, A.J. Ritchie, H.D. (1981). Bailey & Love's Short Practice Of Surgery. London: H.K. Lewis & Co Ltd.

10. Havard School Of Public Health; Vegetable and Fruits. Get Plenty Every Day

11. National Digestive Diseases Information Clearing House (NDDIC)

12. Patient.co.uk. Trusted Medical Information & Support

13. Livingstrong.com. Vitamins & Supplements

14. Wikipedia.en.wikipedia.org/wiki/Dietary- fibre; Gastrocolic Reflex, Toilet

15. Emptyingthe bowel.com. The SquattLooStool

16. Karle, E. Miller, M.D. et al (2000). The Geriatric Patient; Approach to Maintaining Health. American Family Physician 2000 Feb 15; 61 (4) 1089-1104.

17. Lacano, G. et al. New England Journal Of Medicine (1998). Soya Milk. Decreased Chronic Constipation in Young Children.

18. Obach, R.S. Inhibition of Human Cytochrome P450 enzymes by Constituents of St. John's Wort, an herbal Preparation used in Treatment of Depression. J. Pharmacol Exp Ther 2000; 294: 88-95

19. Weber, W. Vander Stoep, A. McCarty, R.L et al (2008) Hypercom (St John's Wort) for Attention Deficit Hyperactivity Disorder in Children and Adolescents: Randomised Controlled Trial JAMA 2008; 299: 2633-41.

20. Medline Plus. A Service Of the U.S National Library Of Medicine NIH Natinal Institute Of Health.

21. Gray, H. (1998). Gray's Anatomy Descriptive And Surgical. Bristol: Parragon.

22. Hull, M.G.R. Joyce, D.N. (1986). Undergraduate Obstetrics and Gynaecology. Bristol: John Wright & Sons.

23. Sinclair, J.M. et al (eds) (2000). Collins English Dictionary and Thesaurus. Glasgow: HarperCollins Publishers.

24. Al Fallouji, M.A.R. Mcbrien, M.P. (1986). Postgraduate Surgery The Candidate's Guide. Oxford: Heinemann Medical Books.

25. Peters, W. Giles, H.M. (1997). A Colour Atlas of Tropical Medicine & Parasitology. Weert Netherlands: Wolfe Medical Publications Ltd.

26. Bell, D.R. (1981). Lecture Notes On Tropical Medicine. London Edinburgh Boston Melbourne Paris Berlin Vienna: Blackwell Scientic Publications.

27. Lucas, A.O. Giles, H.M. (1990). A new Short Textbook of Preventive Medicine for the Tropics. Sevenoaks Kent Great Britain: Hodder and Stoughton Ltd.

28. Donaldson, R.J. Donaldson, L.J. (1993). Essential Public Health Medicine. Dordrecht Boston London: Kluwer Academic Publishers.

29. Rubenstein, D. Wayne, D. (1986). Lecture notes on Clinical Medicine. Oxford London Edinburgh Boston Palo Alto Mebourne: Blackwell Scientific Publications.

30. Karle, E. Zylstra, R.G. Stanbridge, J.B. (2000). The Geriatric Patient: Systematic Approach to Maintaining Health. (The American Family Physician 2000; 61: 1089-104).

31. Lawton, M.P. Brody, E.M. (1969) Assessment Of Older People: self maintenance, and instrumental activities of daily living. Gerontologist 1969; 9: 179-85

32. Cummings, S.R. Nevitt, M.C. Kidd, S. (1988) Forgetting Falls. The Limited Accuracy of Recall od Fallsin the Elderly. Journal AmericanGeriatric Society 1988; 36: 613-16.

33. Lungmore, M. Wilkinson, I. Turmezei, T. Cheung, C.K. (2007). Oxford Handbook of clinical Medicine. Oxford: Oxford University Press.

34. Hennekens, C.H. Buring, J.E. (1987) Epidemiology in Medicine. Boston/Toronto: Little Brown and Company

35. Eveleigh, D.J. (2008). Privies and Water Closets. Oxford: Shire Publications. ISBN 978-0-7478-0702-5.

36. Smith, V.S. (2007) Clean: a History of Personal Hygiene and Purity. p28. ISBN 978-0-19-929779-5.

37. Service, M.W. (1986). Lecture Notes On Medical Entomology. Oxford London Edinburgh Boston: Blackwell Scientific Publications.

38. Tighe, J.R. Davies, D.R. (1988). Pathology. London Philadelphia Toronto Sydney Tokyo: Bailliere Tindall.

39. Florey, C.V. Burney, P. D'Souza, M. Scrivens, E. West, P. (1983). An Introduction to Community Medicine. Edinburgh London Melbourne New York: Churchill Livingstone.

40. NHS Choices, Your Health, Your Choices.

41. Houston, J.C. Joiner, C.L. (1985) A Short Text Book of Medicine. London. Sydney. Auckland. Toronto: Hodder And Stroughton.

42. Hogwood, B.W. Gunn, L.A. (1984). Policy Analysis for the Real World. Oxford: Oxford University Press.

43. Paragon Book (2000). The Complete Works of William Shakespeare. London: Constable & Robinson Ltd.

44. Isaacson, W. (2008) EINSTEIN His Life and Universe. New York London Toronto Sydney: Simon & Schuster Paperbacks

MNEMONICS AND ABBREVIATIONS

3FEV=FRUITS, FIBRE, FLUIDS, EXERCISE, VEGETABLES.

SHIID=STRAINING, HARD STOOL, INFREQUENCY, INSUFFICIENCY, DIFFICULTY

CISSA=CIGARETTES, SALT, SUGAR, ALCOHOL

DISCHO=DIABETES, ISCHAEMIC HEART DISEASE, STROKES, CANCER, HYPERTENSION, OBESITY.

CNS=CENTRAL NERVOUS SYSTEM

CVS=CARDIOVASCULAR SYSTEM

MOT=Ministry Of Transport (Certificate indicating road worthiness of a car)

IBS=Irritable Bowel Syndrome

BMI=Body Mass Index

COPD=Chronic Obstructive Pulmonary Disease